Kurzfassung

Die vorliegende Arbeit untersucht verschiedene Aspekte der Entwicklung supraleitender Nanodraht-Einzelphotonendetektoren (SNSPD) hin zu großflächigen Detektorarrays: die Atomlagenabscheidung für homogene supraleitende Niobnitrid-Dünnschichten, die Weiterentwicklung eines Frequenzmultiplexverfahrens zur Auslese mehrerer Detektoren mit einer reduzierten Anzahl an Leitungen und eine skalierbare Methode zur effizienten optischen Einkopplung.

Die Detektion einzelner Photonen ermöglicht das Erfassen elektromagnetischer Strahlung am Quantenlimit und den vielfältigen Einsatz in Anwendungen wie der Astronomie, der Quantenoptik, der optischen Kommunikation oder der Spektroskopie. Einzelphotonendetektoren basierend auf Supraleitern haben vor allem aufgrund ihrer geringen Energielücke von wenigen meV Vorteile gegenüber Alternativen auf Halbleiterbasis, müssen jedoch auf wenige Kelvin abgekühlt werden. Zu dieser Detektorart gehören SNSPD, welche aus einem supraleitenden Streifen ("Nanodraht") mit einem typischen Querschnitt von 5 nm × 100 nm bestehen. Bei einer Temperatur weit unterhalb der Sprungtemperatur und im Betrieb mit einem Biasstrom nahe des kritischen Stromes reicht die Energie eines einzelnen absorbierten Photons aus, um den supraleitenden Zustand über den gesamten Querschnitt des Nanodrahts zu unterdrücken. Der resultierende endliche Widerstand des Detektors kann als Spannungspuls mit einer typischen Abfallzeit im Nanosekundenbereich erfasst werden. Verglichen mit anderen Einzelphotonendetektoren zeichnen sich SNSPD durch eine hohe Detektionseffizienz bis in den Infrarotbereich, eine hohe Zählrate, eine große Zeitauflösung und eine geringe Dunkelzählrate aus.

Die meisten dieser Charakteristika wurden allerdings mit speziell konzipierten, einzelnen Detektoren ermöglicht, die für viele praktische Einsätze ungeeignet sind. Ein wesentlicher Teil der SNSPD-Entwicklung besteht daher in der Erweiterung hin zu Detektorarrays, möglichst ohne Abstriche hinsichtlich der Detektionseigenschaften machen zu müssen. Die Skalierung einzelner Detektoren hin zu großflächigen Arrays bringt jedoch eine Vielzahl an Schwierigkeiten mit sich, sowohl im Hinblick auf die Fertigungstechnologie, als auch auf die Auslesearchitektur. So werden zur Herstellung der Detektoren supraleitende Dünnschichten mit einer homogenen Schichtdicke und einheitlichen Materialparametern benötigt, um den

Betrieb mit einem Biasstrom nahe des theoretischen Limits zu ermöglichen. In diesem Bereich führt ein Hotspot mit unterdrückter Supraleitung nach Photonenabsorption leichter zur Expansion auf den gesamten Querschnitt des Nanodrahts, was eine höhere Sensitivität und eine verbesserte Zeitauflösung zur Folge hat. Einschnürungen oder Schwachstellen entlang des Nanodrahts limitieren den maximalen Biasstromes der gesamten Struktur und führen daher zu einer Verminderung der Detektoreigenschaften. Ein weiterer wichtiger Aspekt in der Entwicklung von SNSPD-Arrays ist die gemeinsame Auslese mehrerer Pixel durch ein geeignetes Multiplexverfahren, da eine einzelne Auslese jedes Detektors mit einer separaten Leitung durch die Kühlleistung des verwendeten Kryostaten limitiert wird. Eines der zahlreichen vorgestellten Methoden ist das Frequenzmultiplexverfahren, was im Gegensatz zu den meisten Alternativen eine eindeutige Zuordnung der jeweiligen Pixel bei Multiphotonen-Ereignissen erlaubt und die maximalen Zählrate kaum einschränkt, jedoch Abstriche bei der Detektionseffizienz mit sich bringt und aufwendiger auszulesen ist. Zuletzt stellt die effektive optische Einkopplung von Photonen aufgrund der kleinen Abmessungen von SNSPD ein Problem dar. Eine Vergrößerung der aktiven Fläche geht mit einer verringerten Zählrate und einer größeren Defektanfälligkeit in der Fabrikation einher. Andere Einschränkungen der aktiven Fläche werden durch das verwendete Multiplexverfahren auferlegt. Dies trifft vor allem auf das Frequenzmultiplexverfahren zu, da ein großer Flächenanteil von der Resonatorstruktur eingenommen wird, die unempfindlich auf Photonenabsorption reagiert.

Entsprechend dieser Herausforderungen ist die vorliegende Arbeit in die Bereiche Dünnschichtabscheidung, Frequenzmultiplexauslese und optische Einkopplung gegliedert.

Um homogene Dünnschichten zu ermöglichen, wurde ein existierendes Verfahren zur Atomlagenabscheidung von Niobnitrid als Alternative zu Magnetronsputtern im Hinblick auf die Eignung für SNSPD-Arrays untersucht. Messungen an Proben mit Schichtdicken im Bereich (3–21) nm weisen auf eine stark gestörte supraleitende Schicht mit einer kleinen Diffusionskonstante der Elektronen und einem großen spezifischen Widerstand im normalleitenden Bereich (entsprechend einer geringen Elektronenzustandsdichte) hin. Zusammen mit einem kritischen Strom nahe des theoretischen Limits begünstigen diese Eigenschaften die Photonendetektion, da die absorbierte Energie auf ein kleineres Volumen und weniger Quasiteilchen beschränkt bleibt und die Supraleitung dadurch stärker unterdrückt wird. Homogenitätsmessungen an Schichten verteilt über den ALD-Probentisch zeigten nahezu konstante Werte für die Schichtdicke, Oberflächenrauheit, Materialzusammensetzung und elektrischen Parameter über der Fläche eines 2-Zoll-Wafers (50 mm Durchmesser). Für NbN-Schichten mit einem Abstand ≥25 mm zum Zentrum des Probentisches wurde eine Veränderung der Kristallstruktur und eine kleinere Gitterkonstante festgestellt, welche die gemessene Degradation der elektrischen Parameter in diesem Bereich erklären könnten. Aus ALD-NbN-Schichten gefertigte SNSPD zeigten die Charakteristika einer Einzelphotonendetektion und intrinsische

Detektionseffizienzen nahe 100 % für Photonen der Wellenlängen 385 nm und 658 nm. Bei einer Temperatur von 4,2 K wurden Grenzwellenlängen von bis zu 1,25 µm gemessen, welche durch Verringerung der Temperatur noch erhöht werden könnten.

Zur parallelen Auslese mehrerer SNSPD wurde das Frequenzmultiplexverfahren untersucht. Der oszillierende Biasstrom führt zu einem stabilen Detektorbetrieb und einer zuverlässigen Rückkehr des supraleitenden Zustandes nach jedem Detektionsereignis. Die intrinsische Detektionseffizienz folgt exakt dem oszillierenden Biasstrom, wodurch der Detektor aktive und inaktive Zeitfenster mit einer Frequenz im GHz-Bereich aufweist. Dies wurde für eine synchrone Detektion ausgenutzt, für die der Detektor mit einer amplitudenmodulierten VCSEL-Diode bei 3,8 GHz synchronisiert wurde. Für gepulste Anwendungen, bei denen die Zeitspanne eintreffender Photonen bekannt ist, kann dadurch das Signal-zu-Rausch-Verhältnis deutlich vergrößert werden. Im Fall von Photonen mit zufällig verteilten Ankunftszeiten führt der oszillierende Biasstrom allerdings zu einer verringerten durchschnittlichen Detektionseffizienz, im untersuchten Fall von 30 %. Zur Auslese eines Arrays wurde eine Methode mittels Korrelationsfilter beispielhaft an einem Fünf-Pixel-Array untersucht. Durch Korrelation des gemeinsamen Detektorsignals mit zuvor aufgenommenen Pulsen jedes Pixels konnte die eindeutige Zuordnung des ausgelösten Pixels demonstriert und damit eine Ortsauflösung der Zählereignisse erzielt werden. Eine mit diesem Prinzip aufgenommene Statistik kohärenter Photonen zeigte die zu erwartende Poisson-Verteilung.

Möglichkeiten zur effizienten optischen Einkopplung wurden anhand der Integration von SNSPD mit Polymerstrukturen untersucht, die mit direkt-schreibender Laserlithographie hergestellt wurden. Für eine effiziente Freistrahleinkopplung wurden Mikrolinsen auf den aktiven Flächen eines Detektorarrays hergestellt. Diese fokussieren einfallende Photonen auf die Detektoren und vergrößern die effektive Fläche um ein Vielfaches, ohne dabei längere Nanodrähte zu erfordern. Dadurch konnte eine um den Faktor 245 erhöhte Zählrate gegenüber eines Referenzdetektors des Arrays ohne Linse erzielt werden, was dem Verhältnis zwischen beiden effektiven Flächen entspricht. Diese Methode lässt sich leicht skalieren und für verschiedene Array-Anordnungen und -größen anpassen. Für fasergekoppelte Anwendungen wurden SNSPD über photonische Wirebonds direkt mit den Faserkernen eines optischen Faserarrays verbunden. Die Verbindungen waren über mehrere Abkühlvorgänge stabil. Eine Degradation der Nanodrähte beim Schreiben der Wellenleiter konnte durch Aussparungen direkt oberhalb der Detektoren verhindert werden. Verschiedene Methoden wurden eingesetzt, um die optische Koppeleffizienz zu bestimmen. Da eine direkte Messung aufgrund der kleinen intrinsischen Detektionseffizienz bei der verwendeten Wellenlänge von 1550 nm bei 4,2 K nicht möglich war, konnte nur eine grobe Abschätzung über die Abnahme des kritischen Stromes mit Erhöhung der optischen Leistung vorgenommen werden, was zu einer Koppeleffizienz von bis zu 50 % führte.

Ergänzend zu diesen Untersuchungen zur SNSPD-Skalierung wird außerdem eine Methode vorgestellt, um einen der wichtigsten Parameter von supraleitenden Nanodrähten, das Verhältnis des kritischen Stromes zum theoretischen Maximum, zu bestimmen. Diese Methode nutzt die stromabhängige kinetische Induktivität eines Nanodrahts, deren Änderung über eine Resonatorstruktur ausgelesen wird. Im Vergleich zu einer existierenden Methode hat das vorgestellte Prinzip vor allem den Vorteil eines einfacheren Messaufbaus mit integrierter Kalibrierung.

Contents

1 Introduction

Detecting single photons, the quanta of the electromagnetic field, is of great significance in science and engineering. Aside from allowing to detect electromagnetic radiation with the highest, quantum-limited sensitivity, it also enables numerous experiments in fundamental physics. Because of its universal quality, there is a wide range of applications for single-photon detectors in fields like optical communication, astronomy, spectroscopy, ranging, radiometry, and medical imaging [1].

The first device able to detect single-photons was the photomultiplier tube (PMT) which takes advantage of the photoelectric effect and electron multiplication. Nowadays, semiconductor-based single-photon avalanche diodes (SPAD) for the visible and near-infrared spectrum are established as well. They are easy to operate and require, if any, only modest cooling. However, SPADs suffer from high dark-count rates in a trade-off with the detection efficiency [2, 3] and, most importantly, a limited bandwidth due to the semiconductors' comparatively large bandgap on the order of 1 eV at room temperature. Single-photon detectors based on superconductors do not have this limitation because of the much smaller energy gap of only few meV (although not directly comparable because of different detection mechanisms). For their operation, however, they have to be cooled below or close to the critical temperature, typically in the range of few Kelvin.

One of the most promising detectors in this category, the superconducting nanowire single-photon detector (SNSPD), is based on a current-carrying superconducting nanostripe [4, 5]. When biased close to its switching current, the absorption of a single photon is sufficient to drive a part of the nanostripe into the normal state, which can be registered as a voltage pulse across the detector. The SNSPD response offers no inherent energy resolution but is, at the same time, largely insensitive to readout noise. Due to continued efforts in detector development, SNSPDs have been demonstrated with count rates above 2 GHz [6], system detection efficiencies of ≈98 % [7], a timing jitter below 3 ps [8], and dark-count rates below one per hour [9]. Another key feature of SNSPDs is their suitability for wavelengths far beyond the visible range [10, 11]. In particular, detecting photons in the mid-infrared range (3 to 10) µm is of interest for novel applications, e.g., in astronomy [12] or molecular science [13, 14], but is not covered by other types of single-photon detectors. Owing to their exceptional performance metrics, SNSPDs are in use or considered for a wide range of applications like free-space optical communication [15], laser ranging [16, 17], spectroscopy [18], dark-matter search [19], and quantum information processing [20, 21], including quantum key distribution [22] and quantum teleportation [23–25].

However, demonstrations of record metrics have mostly been achieved by specially designed, single-pixel devices. Further SNSPD development is therefore mainly targeted at the combination of all these properties in a single device and the extension to detector arrays with as few compromises as possible. Scaling SNSPDs from single pixels to arrays would enable a variety of new applications in spectroscopy and imaging. The step to large-area SNSPD arrays brings challenges in all fields of detector development. For detector fabrication, one of the most important properties is the uniformity of the superconducting thin film. This applies to small detectors but becomes even more important for large-area devices. Achieving uniform thin films with beneficial properties regarding the detection mechanism of SNSPDs is therefore the subject of continuous material research and deposition optimization. Another challenge is the readout of detector arrays. Operating only a few detectors by a set of high-frequency cables between the cryogenic stage and the room-temperature electronics is entirely feasible. For a larger number of pixels, however, there are constraints in terms of the heat load induced on the cryogenic system, which necessitates multiplexing of multiple detectors to reduce the amount of cables. Several multiplexing architectures have been presented for SNSPDs (see overview in Ref. [26]), with a record number of 1024 pixels shown to date [27]. Nevertheless, as each multiplexing scheme has a different set of advantages and drawbacks, further development effort in all of the various schemes is needed, especially since scaling beyond the current record will likely require the combination of multiple schemes. Yet another challenge in SNSPD development is the rather small active area compared to many other detector types. Typically, the nanowire is meandered over an area with a side length of few µm to several 10 µm. Larger areas require longer nanowires, which go along with worse timing characteristics and an increased susceptibility to fabrication defects. The small active area covered by the nanowire complicates efficient coupling of photons and can severely limit the overall detection efficiency of the detection system. Furthermore, the inter-pixel circuitry of most multiplexing schemes reduces the fill factor considerably. With regard to detector arrays, optical coupling schemes need to be compact, scalable, and flexible to be applicable for different kinds of detector designs. This applies to both free-space and fiber coupling. While multiple coupling schemes with high coupling efficiencies have been demonstrated, most of these methods either require active alignment or are too bulky for arrays.

In view of these technological challenges for scaling up SNSPDs, the aim of this work is the detector development in three main areas:

- *thin-film deposition*, by employing atomic layer deposition as a method to grow large-area uniform superconducting thin films suitable for SNSPDs;
- *multiplexing*, by investigating a previously demonstrated frequency-division multiplexing scheme for SNSPDs to mitigate its main drawbacks;
- *optical coupling*, by integrating SNSPDs with optical elements – microlenses and photonic wire bonds – on top of the active area to improve free-space and fiber coupling to the nanowire.

The basic detection mechanism of SNSPDs is introduced in Chapter 2. Key aspects to consider in SNSPD development are outlined as a basis for the following chapters. The most important metrics to assess the detector performance and a status quo of SNSPD development are given. Different materials suitable for SNSPDs are reviewed, and NbN as the material of choice within this work is motivated. This is followed by an overview of multiplexing architectures for the bias and readout of SNSPD arrays, with a focus on the basic principles of frequency-domain multiplexing and its limitations. The chapter concludes with a summary of existing optical coupling schemes, divided into free-space, fiber-based, and on-chip coupling, in order to motivate the integration with free-form polymer structures within this work.

Chapters 3 and 4 deal with atomic-layer-deposited NbN as a prospective material for SNSPDs, potentially enabling large-area, uniform NbN thin films. Because ALD-NbN has not been used for SNSPD before, basic film and nanobridge properties are investigated first. Afterwards, the film uniformity over the area of a 6-inch wafer is evaluated by measuring the spread in morphology, composition, and superconducting properties. In Chapter 4, first SNSPDs from ALD-NbN are characterized optically regarding the detector response dependent on the applied bias, incoming photon rate, and photon wavelength.

The second area of detector development, frequency-division multiplexing of SNSPDs, is the subject of Chapter 5. This technique was introduced in Ref. [28], and is analyzed further within this work, focusing on the implications of the oscillating bias current on the detector operation and duty cycle. Based on these results, a gated detector operation synchronized to a GHz-modulated laser diode is examined to improve the signal-to-noise ratio in applications with known photon-arrival time spans. Lastly, an alternative readout for frequency-multiplexed SNSPDs based on direct digitization of the detector signal and matched filters to discern between pulses of different pixels, thereby enabling spatial resolution, is proposed.

In Chapter 6, the integration of SNSPDs with free-form optical elements fabricated by three-dimensional direct laser writing is investigated. This technology is already commonly in use, e.g., in silicon photonics, which is why the focus within this work lies on the challenges that come with the integration with SNSPDs operating at cryogenic temperatures. The chapter is divided into two parts, targeting free-space coupling with microlenses and fiber coupling with photonic wire bonds.

As a complement, a method to characterize the quality of superconducting nanowires and extract one of the most important metrics for SNSPDs, the switching to depairing current ratio, is presented in Chapter 7.

2 Superconducting nanowire single-photon detectors

Superconducting nanowire single-photon detectors (SNSPDs) were first demonstrated in 2001 [4, 5] and have since been under extensive research and optimization, exhibiting several excellent detection metrics. This chapter gives an overview of the operation principle and underlying mechanism of SNSPDs. After a short overview of superconducting detectors, the most recent and relevant SNSPD detection models are described in view of their implications on the detector development. In the third part, the detection metrics to assess the suitability for specific applications are given. The most prominently used materials and their compromises are reviewed next. In Section 2.5, various operation schemes for SNSPD arrays are outlined and the concept of frequency-multiplexed SNSPDs is introduced. Lastly, state-of-the-art concepts to efficiently couple light onto the detector's active area are summarized.

2.1 Superconducting detectors

In many elemental and compound materials, electrons can form Cooper pairs when cooled below a certain critical temperature. In this superconducting state, direct current (DC) can be conducted without resistance and magnetic fields are expelled from the superconductor by screening currents within the London penetration depth (perfect diamagnetism).[1] The complex surface impedance in this state can be described in frame of the two-fluid model as

$$Z_s = R_s\left(n_s[T]\right) + i\omega L_s\left(n_s[T], I\right), \tag{2.1}$$

where R_s is the surface resistance and ωL_s is the surface reactance [32]. Factors n_s, T and I denote the Cooper-pair density, temperature, and current, respectively. The "two fluids" in this model, Cooper pairs and unpaired electrons, reflect the super- and normalconducting

[1] A third ramification of the superconducting state are the Josephson effects, used in voltage standards [29], superconducting qubits [30], or the 'superconducting quantum interference device' (SQUID) [31].

components. Using Eq. 2.1, the basic principles of several superconducting detectors can be derived. Transition-edge sensors (TES, $R_s[T]$), kinetic-inductance bolometers (KIB, $L_s[T]$), and microwave kinetic-inductance detectors (MKID, $L_s[n_s]$) all employ different mechanisms to disturb the superconducting state for photon detection. SNSPDs, the main topic of this thesis, are based on a current-assisted breakdown of the superconducting state which involves the Cooper-pair-density dependence $R_s(n_s)$ in the detection process. Apart from photon detectors, the dependence $L_s(I)$ is employed in "microwave kinetic-inductance nanowire galvanometers" (M-KING), originally shown as current sensors [33], but investigated here in Chapter 7 as a characterization method for nanowires.

2.2 Operation principle of SNSPDs

Superconducting nanowire single-photon detectors consist of a superconducting wire with a typical thickness $\leq$10 nm and a width on the order of (20...200) nm. This nanowire is cooled well below its critical temperature (usually $\lesssim 0.5T_c$) and is biased by a current close to the switching current ($I_b \lesssim I_{sw}$). An absorbed photon creates excited quasiparticles forming a confined region ("hotspot") with a suppressed superconducting order parameter. Depending on the operation conditions, this hotspot either relaxes again or – by the entry and movement of vortices – expands into a normal domain spanning the whole width of the nanowire. Because of this transition to the resistive state, the applied bias current is redirected out of the nanowire into the readout electronics, which, in turn, leads to a measurable voltage pulse. In the absence of a current through the nanowire, the normal domain shrinks and the superconducting state gets restored. A sketch of the SNSPD detection principle is shown in Fig. 2.1.

In the following, the detection process is described in more detail by means of several comprehensive works published in recent years. These models of the detection mechanism are not only important for a better understanding of the detection process itself, but also for practical detector development. Knowing the factors which influence the detector's performance is necessary for choosing suitable materials, optimizing detector layouts, and conceiving novel operation schemes.

2.2.1 Hotspot formation

In the initial superconducting state and $0 < T < T_c$, both Cooper pairs and thermally excited quasiparticles (single electrons or holes) exist in the superconductor. The excited states are

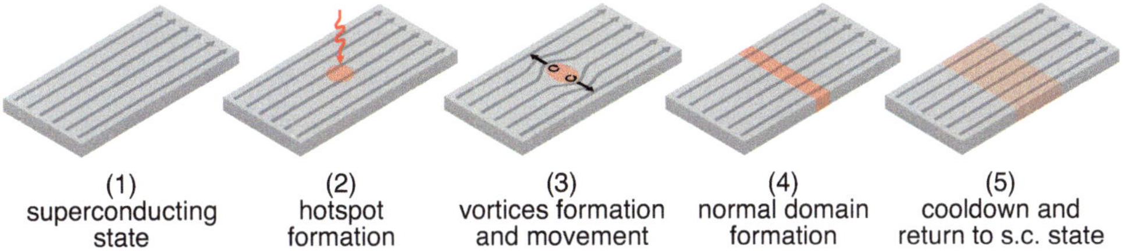

Figure 2.1: Basic photon detection process in current-biased nanowires. Initially, the nanowire is in the current-carrying, superconducting state (1). Upon photon absorption, Cooper pairs get excited above the superconducting gap and create elevated quasiparticles and phonons, which form a hotspot (2). Inside the hotspot, the superconducting order parameter is suppressed and vortices can form. Because of the bias current, the Lorentz force acts on the vortices which dissipatively move transversely across the nanowire. This leads to a normal domain over the whole width (4), with a resistance in kΩ-range. Thus, the bias current is redirected from the nanowire to the readout circuit which can be detected as a voltage pulse over the load resistance. Without a current through the nanowire, the normal domain cools down and the initial superconducting state gets restored (5).

separated from the superconducting ground state of the Cooper pairs by the temperature-dependent energy gap

$$\Delta(T) = \Delta(0)\left[1 - \left(\frac{T}{T_{\mathrm{c}}}\right)^2\right]^{\frac{1}{2}}\left[1 + \left(\frac{T}{T_{\mathrm{c}}}\right)^2\right]^{\frac{3}{10}} \tag{2.2}$$

with the zero-temperature energy gap $\Delta(0) = 1.76k_{\mathrm{B}}T_{\mathrm{c}}$ [34, 35][2] and Boltzmann constant k_{B}. Equation 2.2 is an approximation to an implicit temperature dependence following from the BCS theory (see Eq. 6.35 in Ref. [37]). In the typical operation range of SNSPDs ($T \lesssim 0.5T_{\mathrm{c}}$), $\Delta(T) \approx \Delta(0)$ and Cooper-pair density $n_{\mathrm{s}}(T) \approx n_{\mathrm{s}}(0)$.

In order to break such a pair, photons need an energy $\hbar\omega \geq 2\Delta(T)$ (factor 2 because the pair breaks into two unbound electrons). For an exemplary value of $T_{\mathrm{c}} = 10\,\mathrm{K}$, $2\Delta(4.2\,\mathrm{K}) \approx 2.9\,\mathrm{meV}$. Although this allows for photons in the far-infrared range to break a single Cooper pair, the subsequent steps involved in the detection process in practice require much higher energies to actually generate a detectable pulse.

When a photon of higher energy (in optical or near-infrared range) is absorbed, the generated quasiparticles can get excited several orders of magnitude above the superconducting energy gap. In an energy down-conversion process, the high-energy quasiparticles interact via

[2] The ratio $\beta_0 = 1.76$ between $\Delta(0)$ and $k_{\mathrm{B}}T_{\mathrm{c}}$ applies to weakly coupled superconductors in the BCS theory. For strongly coupled superconductors like niobium nitride, β_0 was experimentally found to be higher at ≈ 2.05 [36].

electron-electron scattering (with characteristic time τ_{ee}), electron-phonon scattering (τ_{ep}), and phonon-electron scattering (τ_{pe}), creating excited quasiparticles and hot phonons until thermalization of the electronic and phonon system at temperatures T_e and T_{ph} is reached. This typically happens on a timescale of few picoseconds after photon absorption (see Fig. 2 in Ref. [38]).

A comprehensive study of the detection mechanism of SNSPDs was presented by Vodolazov in 2017 [38]. To facilitate modeling the whole detection process, Vodolazov splits up the process in a timescale up to the characteristic time of variation in the order parameter ($t \lesssim \tau_{|\Delta|} = \hbar/k_B T_c$) – at which the order parameter is suppressed and the initial hotspot formed – and beyond this time. For $t \lesssim \tau_{|\Delta|}$, the evolution of the electron and phonon distributions after photon absorption are modeled by kinetic equations (Eqs. 1 and 2 in Ref. [38]).

One of the main results of this model is the importance of the ratio of the electron and phonon heat capacities at the critical temperature of the used superconductor, $C_e/C_{ph}\big|_{T_c}$. This ratio determines the portion of the absorbed energy going to the electronic system. Larger ratios lead to shorter thermalization times and a stronger suppression of the superconductive state [38]. This is preferred as it facilitates photon detection. A second crucial parameter is the electron diffusion coefficient D. Small values of D also favor a fast thermalization because the diffusion of excited electrons out of the absorption site is slower.

For times $t \gtrsim \tau_{|\Delta|}$, Vodolazov considers two cases, depending on whether the electrons are already fully thermalized at $\tau_{|\Delta|}$ or not [38]. For the limit of a very long thermalization time comparable to the electron diffusion time to reach the nanowire edges ($\tau_{th} \approx \tau_{D,w} = w^2/4D$, with nanowire width w), a "hot-belt model" is assumed (see Fig. 4 in Ref. [38]). This case, however, is considered to be irrelevant for materials typically used for SNSPDs.

In the more appropriate case, applicable to superconducting materials with a small coefficient D and a large ratio $C_e/C_{ph}\big|_{T_c}$, the thermalization time is on the order of or smaller than the characteristic time of order parameter variation ($\tau_{th} \lesssim \tau_{|\Delta|}$). Therefore, electrons and phonons are considered to be completely thermalized at time $\tau_{|\Delta|}$ and the radius of the initial hotspot is on the order of the coherence length ξ. This is used by Vodolazov as a starting point for modeling the evolution from the generated hotspot onward.

2.2.2 Detection mechanism

Although the exact hotspot formation process was not yet clear to the degree it is today, the general concept of a hotspot that leads to the breakdown of superconductivity inside the nanowire was thought of already in the first publications [4, 5]. The mechanism that determines the evolution from a local hotspot to the resistive state of the nanowire, however, was not as established and the subject of extensive research, leading to various models over time (see overview in Ref. [39]): normal-core hotspot model [4, 5], diffusive hotspot model [40], vortex-entry model [41, 42], diffusion-based vortex entry model [43, 44], and normal-core vortex model [45–48].

Within the course of understanding the underlying mechanism of the SNSPD response, the critical role of vortices for the breakdown of the superconducting state has become more and more apparent. Vortices, which can exist in type-II superconductors, are spatially confined areas where the magnetic field can penetrate the material. Each vortex carries a flux quantum $\Phi_0 = h/(2e)$ with the magnetic field decaying from the core with a distance of the London penetration depth λ_l, and has a vanishing Cooper-pair density n_s, varying over the coherence length ξ. By applying a bias current, the vortices are moved by the Lorentz force through the superconductor. This movement can lead to the expansion of the hotspot to a normal domain across the nanowire.

In the work by Vodolazov, the conditions determined by the energy down-conversion cascade (thermalization at elevated temperature $T_e = T_{ph} > T$ inside the hotspot, and $|\Delta|(T)$ depending on this temperature) serve as a starting point for simulations of the hotspot dynamics at $t > \tau_{|\Delta|}$. Using equations modeling the dynamics of the electron and phonon temperatures (Eq. 30 and 31 in Ref. [38]), and a modified, temperature-dependent Ginzburg-Landau equation (Eq. 36 in Ref. [38]), the creation and expansion of the normal domain inside the nanowire is simulated.

This expansion can be traced back to the nucleation of vortices near or inside the hotspot with suppressed $|\Delta|$ in the presence of a bias current. Depending on the location of the hotspot across the width of the nanowire, different cases can be distinguished [38]. If the hotspot is located near the edge of the nanowire, single vortices can enter from this edge with reduced $|\Delta|$ and subsequently move to the opposite edge of the nanowire due to the bias current and resulting Lorentz force. In the case of a hotspot located toward the center of the nanowire, the nucleation of a vortex-antivortex pair (VAP) inside the hotspot is possible, which then move to opposite edges of the nanowire.

If the bias current is large enough, i.e., larger than a threshold called detection current I_{det}, the dissipative vortex movement leads to a growing normal domain across the whole width of the nanowire. The detection current $I_{det}(y, E_{ph}, w, T, H)$ depends on the position y across the nanowire but also on the photon energy E_{ph}, the nanowire width w, the operating temperature T, and the magnetic field H.

The detection current can vary in a range between the retrapping current I_r (below which the cooling of the nanowire exceeds the Joule heating in the resistive state, thus prohibiting any formation of a normal domain) and the depairing current I_{dep} (the theoretical limit of the current a superconductor can carry, determined by the critical kinetic energy of the Cooper pairs), $I_r < I_{det} < I_{dep}$. For any photon energy E_{ph}, there is a threshold $I_{det} < I_{dep}$ above which the photon could be detected. However, the bias current I_b in the superconducting state is limited by the switching current I_{sw}. In practice, the critical current $I_c < I_{dep}$ because of nonuniformities and defects in the film. In addition, quantum and thermal fluctuations can cause a switching to the normal state at a lower, experimentally determined current $I_{sw} < I_c$. If $I_{sw} < I_{det}$, the nanowire cannot be biased above the detection current and a hotspot, formed after energy absorption, does not lead to a resistive state of the nanowire. The different current parameters can be summarized as

$$0 \le I_r \le I_{sw} \le I_c \le I_{dep}. \tag{2.3}$$
$$\quad\quad \longleftarrow I_{det} \longrightarrow \text{ (from } I_r \text{ to } I_{dep}\text{)}$$
$$\longleftarrow I_b \longrightarrow \text{ (from } 0 \text{ to } I_{sw}\text{)}$$

As a result, an important property of superconducting nanowires is the ratio I_{sw}/I_{dep} because a higher value of this ratio goes along with the ability to detect photons of significantly smaller energies (toward mid-infrared range) or, conversely, relaxes the requirements for the operating temperature and the nanowire width at a given photon energy. For ratios $I_{sw}/I_{dep} > 0.7$, Vodolazov has predicted micrometer-wide stripes to be able to detect near-infrared or optical photons (see near-overlapping curves above $I_{det}^{max}/I_{dep} = 0.7$ in Fig. 11, Ref. [38]). This was later realized and experimentally confirmed by Korneeva *et al.* [49].

2.2.3 Response dynamics

Due to Joule heating, the normal domain grows along the nanowire, leading to a resistance R_n on the order of few kΩ [50]. This resistance ($\gg 50\,\Omega$) causes the bias current to divert out of the nanowire into the load resistance of the readout circuit (typically $R_l = 50\,\Omega$), see basic circuit diagram of the SNSPD operation in Fig. 2.2a. The resulting pulse height can be estimated by $I_b \cdot R_l$ and typically is on the order of (0.1 to 10) mV before amplification.

After a detection event, at a current $0 \le I < I_r$ through the nanowire, the superconducting state gets restored, the bias current returns back to the nanowire, and the measured voltage over the load inductance vanishes.

The timescales of these response dynamics are mainly governed by the nanowire's inductance L. The total inductance $L = L_m + L_k$ is made up by the magnetic inductance L_m, which is

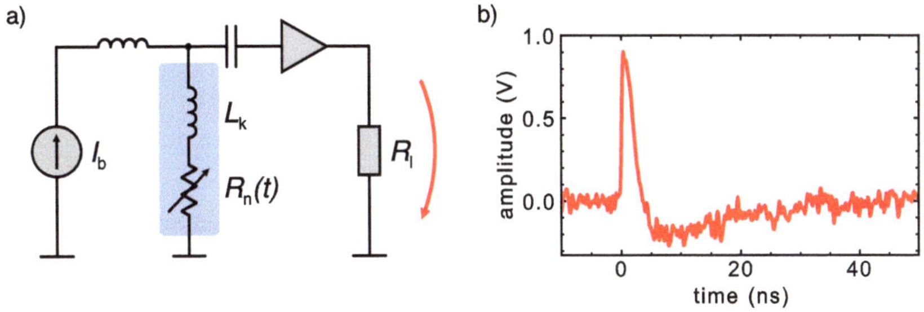

Figure 2.2: Basic SNSPD circuit diagram and detector pulse. (a) The SNSPD, marked in blue, is modeled by a kinetic inductance L_k and a time-dependent resistance $R_\mathrm{n}(t)$ in series. The detector is biased by a current source through the DC/inductive port of a bias tee, and read out by the RF/capacitive port and an amplifier. A detection event is seen as a sudden rise of $R_\mathrm{n}(t)$ from $0\,\Omega$ to $\approx$kΩ, causing the bias current to divert into the load resistance R_l. (b) A typical, amplified SNSPD pulse with a rise time τ_rise of (10 to 100) ps and an exponential fall time τ_fall of few ns.

the result of energy stored in the magnetic field in and around the superconductor, and a kinetic inductance L_k, which comes from the inertia of the Cooper pairs.[3] At typical operation conditions and detector geometries, $L_\mathrm{k} \gg L_\mathrm{m}$ and L_m can be neglected.

The kinetic inductance can be calculated as

$$L_\mathrm{k} = \frac{m_\mathrm{e}}{n_\mathrm{s}(T)e^2}\frac{l}{wd} = \mu_0\lambda_\mathrm{l}^2(T)\frac{l}{wd} \approx \frac{\hbar R}{\pi\Delta(0)}, \tag{2.4}$$

with the London penetration depth

$$\lambda_\mathrm{l}(T) \underset{T\lesssim 0.5T_\mathrm{c}}{\approx} \lambda_\mathrm{l}(0) = \sqrt{\frac{\hbar\rho_\mathrm{n}}{\pi\mu_0\Delta(0)}}, \tag{2.5}$$

electron mass m_e, elementary charge e, vacuum permeability μ_0, residual resistivity ρ_n, and geometry parameters l, w, and d (see Eqs. 2.21 and 2.51 in Ref. [37] and p. 163 f. in Ref. [51]).

With L_k, the pulse rises with a time $\tau_\mathrm{rise} = L_\mathrm{k}/\,(R_\mathrm{l} + R_\mathrm{n})$ (equivalent to the decay time of the current through the nanowire) and fall time $\tau_\mathrm{fall} = L_\mathrm{k}/R_\mathrm{l}$ [52]. Because the normal-state resistance of the nanowire $R_\mathrm{n} \gg R_\mathrm{l} = 50\,\Omega$, the fall time is several orders of magnitude larger than the rise time. For a typical square kinetic inductance $L_\mathrm{k,sq} = 50$ pH and SNSPD dimensions $w = 100$ nm and $l = 100\,\mu$m, the pulse fall time amounts to $\tau_\mathrm{fall} \approx 1$ ns.

[3] Because of the short collision time of electrons compared to Cooper pairs, the kinetic fraction of the total inductance in normal conductors can usually be neglected.

The kinetic inductance is directly proportional to the length l of the nanowire. A low kinetic inductance can cause the detector to stay in the normal domain after a detection event. This is called latching [53]. When the nanowire is shorter and therefore its kinetic inductance smaller, the bias current returns back to the nanowire faster after a detection event. If the current returns before the superconducting state in the nanowire is restored, i.e., when the hotspot lifetime exceeds the reset time, the detector stays in a stable, self-heating, resistive state and no photon detection is possible. Therefore, the maximum count rate is limited by the thermal relaxation of the hotspot. Common materials have hotspot relaxation times in the range (10 to 500) ps, depending on the bias current, operating temperature and induced photon energy [54–57].

Several electrothermal models have been developed to describe the reset dynamics of SNSPDs [50, 53, 58, 59]. Berggren *et al.* showed that the response dynamics can be modeled in SPICE[4] [60], greatly simplifying the simulation of SNSPD circuits. This was later modified by McCaughan *et al.* to account for thermal fluctuations [61], which are inherent to SNSPDs and depend exponentially on the bias current through the device.

Using these models, the evolution of the hotspot resistance R_{hs} can be described by (adapted from Eqs. 2 and B.1 in Ref. [60] and Eq. 1 in Ref. [53])

$$\frac{\mathrm{d}R_{\text{hs}}}{\mathrm{d}t} = \frac{R_{\text{sh}}}{w}\frac{\mathrm{d}l_{\text{hs}}}{\mathrm{d}t}, \tag{2.6}$$

with normal-state sheet resistance R_{sh} and the expansion of the hotspot length

$$\frac{\mathrm{d}l_{\text{hs}}}{\mathrm{d}t} = 2v_0 \cdot \frac{\left(\frac{I_{\text{n}}}{I_{\text{sw}}}\right)^2 \psi - 2}{\sqrt{\left[\left(\frac{I_{\text{n}}}{I_{\text{sw}}}\right)^2 \psi - 1 + \left|\left(\frac{I_{\text{n}}}{I_{\text{sw}}}\right)^2 \psi - 1\right|\right]/2 + \delta}} \tag{2.7}$$

$$= 2v_0 \cdot \frac{a-2}{\sqrt{\left[a - 1 + |a-1|\right]/2 + \delta}}, \; a = (I_{\text{n}}/I_{\text{sw}})^2\,\psi. \tag{2.8}$$

Parameter v_0 is the characteristic velocity of hotspot growth, I_{n} is the current through the nanowire, ψ is the Stekly parameter (describing the ratio of Joule heating to heat transfer in the normal state [62]), and $\delta \ll 1$ a small positive offset [60]. The denominator in Eq. 2.7 follows from Eq. 1 in Ref. [53] by taking its real part and adding δ. By this, Berggren *et al.* avoided problems with the term under the square root getting zero or negative (see Appendix B in the supplementary of Ref. [60]). One of the shortcomings of this model is the assumption of a lumped-element representation of the SNSPD like shown in Fig 2.2a.

4 Simulation Program with Integrated Circuit Emphasis.

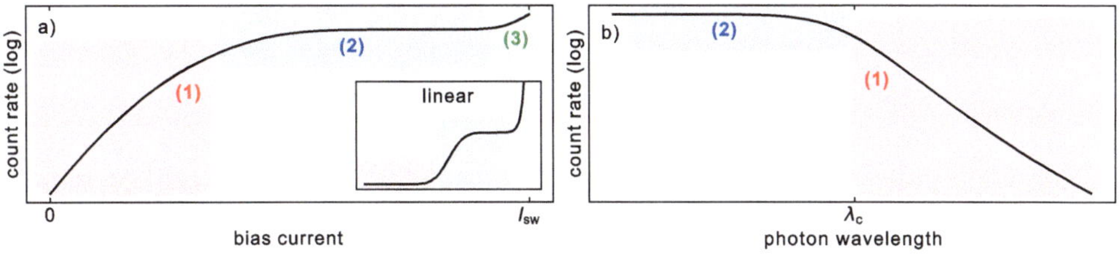

Figure 2.3: Basic dependence of the count rate on the applied bias current at constant λ (a), and on the photon wavelength at constant I_b (b). Dependence CR (I_b) can be modeled by Eqs. 2.10 and 2.11, while PCR (λ) is given by Eq. 2.12. The highlighted areas are the probabilistic regime (1, red), deterministic regime (2, blue), and dark-count-dominated regime (3). In the inset of (a), the same CR(I_b) curve is plotted in linear scale.

For very long nanowires, this assumption is no longer adequate and a distributed element model is required instead [63, 64]. Nevertheless, the phenomenological model by Kerman *et al.* [53] and Berggren *et al.* [60] still has its merits, particularly for its convenience, and has proven to be able to accurately estimate rise and reset times, pulse amplitudes, and the latching behavior of common SNSPDs [65]. Equations 2.6 and 2.7 will be used in Section 5.2 to simulate the reset and latching behavior of SNSPDs with an oscillating bias current.

2.3 Detection metrics

2.3.1 Detection efficiency and spectral dependence

The efficiency to detect a photon can be split up into multiple factors. The total system detection efficiency, that is the registered photon-count rate divided by the number of photons sent to the detection system, can be summarized as

$$\mathrm{SDE} = \eta_{\mathrm{oc}} \cdot \underbrace{\eta_{\mathrm{abs}} \cdot \eta_{\mathrm{int}}}_{\mathrm{DE}} = \frac{R_{\mathrm{col}}}{\mathrm{PR}} \cdot \frac{R_{\mathrm{abs}}}{R_{\mathrm{col}}} \cdot \frac{\mathrm{PCR}}{R_{\mathrm{abs}}} = \frac{\mathrm{PCR}}{\mathrm{PR}}. \tag{2.9}$$

The optical coupling efficiency η_{oc} is the efficiency of the detection system to couple photons with a rate PR onto the active area of the detector, here written as collection rate R_{col}, and will be discussed in Section 2.6.

The absorption efficiency η_{abs} and intrinsic efficiency η_{int} are mainly determined by the detector itself and are usually jointly expressed as detection efficiency DE. The efficiency to absorb (with rate R_{abs}) an impinging photon depends on the photon wavelength and material, specified by the extinction coefficient k, the imaginary part of the complex refractive index $\boldsymbol{n} = n - ik$. In the case of illumination normal to the substrate plane, η_{abs} is mainly limited by the small absorption length given by the film thickness of few nm. For $\lambda = 1550\,\mathrm{nm}$ and $d = 5\,\mathrm{nm}$, η_{abs} amounts to about 30 % (see Fig. 4 in Ref. [66]). To increase the absorption, the nanowire can be embedded in cavities which reflect photons to traverse the detector multiple times. This has been shown by various groups [67–70] and absorption efficiencies close to 1 have been achieved for a small bandwidth around the designed wavelength. Besides normal incidence, photons can also be coupled to the detector by guiding them along the nanowire, using the nonzero evanescent field of the guided mode [71]. This way, the absorption is not limited to the film thickness and absorption lengths >10 µm can be realized.

The intrinsic detection efficiency η_{int} depends on the operation regime determined by the bias current and photon wavelength. Experimentally, this can be extracted from $\mathrm{CR}(I_{\mathrm{b}})$ and $\mathrm{CR}(\lambda)$ measurements. In Fig. 2.3, these dependences are depicted qualitatively. The bias dependence $\mathrm{CR}(I_{\mathrm{b}}) = \mathrm{PCR}(I_{\mathrm{b}}) + \mathrm{DCR}(I_{\mathrm{b}})$ consists of a sigmoidal-shaped photon-count rate [72]

$$\mathrm{PCR}\,(I_{\mathrm{b}}) = \frac{\mathrm{PCR}_{\mathrm{sat}}}{2} \cdot \left(1 - \mathrm{erf}\left[\frac{I_0 - I_{\mathrm{b}}}{\Delta I_{\mathrm{b}}}\right]\right) \tag{2.10}$$

and an exponential dark-count rate [39] which can be expressed as

$$\mathrm{DCR}\,(I_{\mathrm{b}}) = a \cdot e^{I_{\mathrm{b}}/b}. \tag{2.11}$$

Parameter $\mathrm{PCR}_{\mathrm{sat}}$ is the photon-count rate in the saturated regime, "erf" is the error function, I_0 is the current at the inflection point of the curve, and ΔI_{b} is the transition width of the current dependence. Parameters a and b are fitting coefficients of the exponential curve.

The three operation regimes shown in Fig. 2.3a can be distinguished as:

(1) Probabilistic regime: At low currents, the energy of an absorbed photon is not sufficient to trigger a detection event. Instead, the photon can only be detected with the help of, e.g., thermal fluctuations which cause vortex penetration near the hotspot, so-called "fluctuation-assisted vortex entry" [47].

(2) Deterministic regime: If the current is large enough, a photon triggers a switch of the nanowire to the normal state deterministically without the need of any additional fluctuations. In this regime, the count rate is independent of the bias current and the intrinsic detection efficiency η_{int} is assumed to be close to 100 % [47].

(3) Dark-count-dominated regime: In this regime, the current is high enough that vortices can penetrate the nanowire even without the energy of an absorbed photon. These dark counts dominate photon counts at currents close to I_{sw}.

There are several reasons for the fact that the CR(I_b) curve does, in linear scale, not resemble a sharp step function but instead exhibits a certain transition width ΔI_b (see Ref. [72] and summary in Ref. [73]). One reason is the position-dependent detection current $I_{det}(y)$. This threshold varies between I_{det}^{min} and I_{det}^{max} depending on the hotspot location across the strip width (see Fig. 9 in Ref. [74]). If the detector is biased by a current within this range, only part of the nanowire takes part in the detection. Only for $I_b > I_{det}^{max}$, also the least sensitive part (in most cases the center of the nanowire [44, 47, 48]) contributes to photon detection and the whole nanowire is active. Another reason are nonuniformities of the nanowire. These include a spatially varying energy gap Δ [75], constrictions in the nanowire geometry both in width and thickness [76], and the granular structure in polycrystalline films [77]. These imperfections lead to a nanowire which is not biased evenly at a constant ratio I_b/I_{sw} over the whole length. Lastly, Kozorezov *et al.* introduced Fano fluctuations [72] as a reason for the broadening of the error function, even in the absence of material inhomogeneities. These fluctuations occur during the energy down-conversion process after photon absorption and impact the amount of energy delivered to the electronic system.

In Fig. 2.3b, the same operation regimes are highlighted in the wavelength dependence of the count rate. The curve can be described by [78]

$$\mathrm{PCR}\,(\lambda) = \frac{1}{1+\left(\frac{\lambda}{\lambda_c}\right)^p} \tag{2.12}$$

with photon wavelength λ, cut-off wavelength λ_c, and a factor p which determines the slope of the declining curve toward longer wavelengths.

The shape of PCR(λ) is linked to the one from PCR(I_b) and is subject to the same mechanisms leading to a broadened transition around the cut-off. Semenov *et al.* showed that the existing contributions (position dependence, inhomogeneities, Fano fluctuations) are not sufficient to explain experimental data on the spectral broadening of the photon-count rate, and introduced local thermal fluctuations as the main reason instead [73]. These fluctuations are estimated to take effect on a length scale of $\approx$10 nm, which is smaller than the typical hotspot radius.

The cut-off wavelength λ_c (corresponding to a minimal photon energy $E_{ph,min}$) is one of the most important SNSPD metrics, and pushing λ_c toward longer wavelengths (mid-IR range) is a leading aim of detector development. A larger value of λ_c at a constant current ratio I_b/I_{sw} is qualitatively equivalent to a smaller saturation current I_{sat} (commonly defined as the bias current at which 90 % of the saturated photon-count rate is reached) at a constant wavelength.

There are material- and geometry-dependent parameters to increase λ_c. From the detection process described in Sections 2.2.1 and 2.2.2, the diffusion coefficient D and the heat capacity ratio $C_e/C_{ph}\big|_{T_c}$ play a crucial role when trying to extend the photon-detection ability to smaller energies. A small $D \lesssim 0.5\,\mathrm{cm^2/s}$ together with a ratio $C_e/C_{ph}\big|_{T_c} \gg 1$ lead to a fast thermalization of quasiparticles ($\tau_{th} < 1$ ps) and a stronger suppression of the superconducting state at a given photon energy [38]. The energy gap $\Delta \propto T_c$ of the material is another optimization parameter to improve λ_c. Lower-T_c materials have a smaller gap Δ and therefore require a smaller photon energy to suppress superconductivity. A low electron density of states $N(0) = 1/(2e^2\rho_n D)$ (large resistivity ρ_n; Einstein relation) is also beneficial for the detector's energy sensitivity because the absorbed energy of a photon leads to a larger fraction of excited quasiparticles.[5] SNSPD materials and their properties will be discussed in Section 2.4.

In addition to material parameters, the cross-section wd of the nanowire is of great importance to increase the cut-off wavelength. This was already apparent in early models of the detection process [40]. The cut-off wavelength $\lambda_c \propto (wd)^{-1}$ is inversely proportional to the thickness [79] and width [80] of the nanowire. Marsili *et al.* investigated width-dependent detection efficiencies and were able to push the sensitivity of SNSPDs from $\lambda \approx 2\,\mu\mathrm{m}$ at $w = 85$ nm up to $\lambda \approx 5\,\mu\mathrm{m}$ at $w = 30$ nm for the same relative bias current I_b/I_{sw} [10]. When reducing the cross-section, it is important to consider that, because $I_{sw} \propto wd$, this goes along with a smaller signal amplitude and therefore also lower signal-to-noise ratio.

The last and perhaps most important optimization parameter is the current ratio I_{sw}/I_{dep}. It is connected to both the material and geometry of the detector, and typically lies in the range 0.5 to 0.6 [80, 81]. High ratios I_{sw}/I_{dep} allow to bias the detector closer to the depairing current, which can greatly extend the sensitivity to smaller photon energies. As mentioned in Section 2.2.2, it also relaxes the requirements for the cross-section. However, there are multiple factors limiting the switching current. To increase the area covered by the SNSPD, the nanowire is usually designed as a meandered line instead of a straight stripe. This adds bends with angles of 90° or 180° to the nanowire. In the inner boundary of these bends, the current density is higher than in the straight parts of the nanowire, which reduces the switching current of the whole device. This effect, called "current crowding", was modeled by Clem/Berggren [82] and later experimentally investigated on various geometries [83, 84]. Several optimizations of the detector layout are feasible to reduce current crowding. In

5 On the other hand, $C_e \propto N(0)$ and, in general, a smaller $N(0)$ also goes along with a smaller switching current.

Ref. [82], the optimal bend curvature for various layouts was calculated by modeling the energy barrier for vortex entry. Charaev *et al.* investigated spiral nanowires (double-spiral, and single-spiral with a bridge connecting the center) as an alternative to meandered detectors to reduce current crowding and were able to increase I_{sw}/I_{dep} from 0.4 (meander) to 0.48 (double-spiral) and 0.55 (single-spiral) [85]. This has translated to an increase in λ_c from ≈600 nm (meander) to ≈900 nm (single-spiral) at the same I_b/I_{sw}. Recently, regular meandered SNSPDs with a variable thickness were presented [86]. The 180°-turns are about twice as thick as the straight parts of the meander. With this method, I_{sw} could be increased by ≈30 %, an effect attributed to a minimization of current crowding in the bends.

Another potential influence on I_{sw}/I_{dep} comes from constrictions in the cross-section along the nanowire length, which can severely limit the cut-off wavelength [76]. These constrictions (variations in film thickness and/or nanowire width) limit the switching current and therefore cause other parts of the nanowire to be "underbiased". Hence, highly uniform nanowires, both laterally and vertically, are the key goal in the detector fabrication. The film uniformity can be optimized by the choice of the material, deposition method, and deposition parameters when growing the thin films. The substrate material and optional buffer layers are further factors. Similarly, optimizing the patterning process to improve the edge roughness of the nanowires can significantly increase ratio I_{sw}/I_{dep} [87]. This is particularly important for narrower stripes because the roughness of the nanowire edges relative to the width increases.

Aside from layout and fabrication optimizations, the operating temperature has a significant impact on the photon detection capability of SNSPDs. Lowering T in the range below $0.7T_c$ reduces the minimal detectable photon energy at constant I_b/I_{dep} (see Fig. 12 in Ref. [38]). This dependence has also been shown experimentally [54, 69, 78]. Contrary to the idea to operate SNSPDs at temperatures close to T_c where the energy gap sharply decreases (Eq. 2.2) in order to facilitate photon detection, the normal domain cannot grow close to T_c because of the reduced ratio C_e/C_{ph} and the vanishing $I(V)$ hysteresis (at some point $I_r = I_{sw}$) [38].

In terms of overall system detection efficiency, considerable advances have been made over the last decade (see overview in Tab. 1, Ref. [88]). In 2013, for the first time an SDE above 90 % at 1550 nm was demonstrated using a cavity-integrated SNSPD and adjusted polarization of the light [69]. Later, similar SDE values have been achieved by other groups [70, 89, 90]. Considerably higher SDEs at ≈98 % have only been reported recently at 1550 nm [7], 1350 nm [91], and 1590 nm [92].

While most of the SDE optimizations have been concentrated around a wavelength range (1200 to 1700) nm, which is typically used for optical communications and other applications [93], there are also efforts to push the cut-off wavelength into mid-IR. Marsili *et al.* demonstrated single-photon sensitivity up to 5 µm by decreasing the nanowire cross-section [10]. A similar approach was adopted by Korneev *et al.*, showing single-photon detection at 10.6 μm using a 40 nm-wide detector [94]. Verma *et al.* demonstrated straight, 10 μm-long nanowires with a cross-section of 2.61 nm × 50 nm exhibiting an incipient saturation of the PCR(I_b) curve at 9.9 μm wavelength [11].

2.3.2 Dark-count rate

In general, any count registered by the detection system not originating from a photon which should be detected can be considered a dark count, whether it comes from electronic noise, background radiation, or detector-intrinsic causes. In a narrower sense, only the intrinsic dark counts are considered and can be treated as the noise level of the detector at given operation conditions.

Intrinsic dark counts in SNSPDs are predominantly caused by thermally activated vortices [39, 41, 95]. To enter the nanowire, vortices have to overcome an energy barrier. With higher currents, the barrier is reduced and thermally excited vortices can more easily penetrate inside. This happens mainly at weak points along the nanowire, for example at the inner border of bends where the barrier is smaller because of a higher current density (current crowding). The influence of nanowire bends on vortex entry and dark counts has been investigated by applying an external magnetic field [42, 96]. Measurements on square spirals revealed an asymmetric magnetic-field dependence of the dark-count rate (DCR) [97]. All bends of these spirals are oriented in the same direction with respect to the bias current and magnetic field, and by applying a magnetic field, the barriers on the inner border are either increased or further decreased. For bend-free structures (circular single-spirals with a radius $r \gg w$), no such asymmetry was found. These measurements substantiate that dark counts are mainly caused by vortex hopping from weak points in the bends.

Because the entry barrier is smaller for higher currents, the dark-count rate of an SNSPD exponentially increases with the bias current according to Eq. 2.11. Due to being thermally activated, the rate of dark counts also depends on the temperature and decreases for smaller T at constant I_b/I_{sw} (e.g., Fig. 2b in Ref. [71]). Experimental support of this has been reported in Ref. [98], where DCR could be lowered by improving the thermal coupling between the detector substrate and the heat sink.

In general, SNSPDs exhibit very low DCRs compared to other single-photon detectors (Tab. 1 in Refs. [99, 100]), especially if there is a broadly saturated photon-count rate over bias current and the detector can therefore be operated at a bias level well below the switching current. Highly efficient SNSPDs with intrinsic dark-count rates below 1 Hz have been demonstrated [69].

2.3.3 Maximum count rate

The maximum count rate (MCR) is determined by the reset time $\tau_r = \tau_{rise} + \tau_{fall} \approx \tau_{fall} = L_k/R_l$. As described in Section 2.2.3, typical reset times of SNSPDs are on the order of few ns, enabling count rates of several 100 MHz. The load resistance R_l is typically fixed at 50 Ω, therefore MCR can only be increased by a reduction of the kinetic inductance L_k.

As seen from Eq. 2.4, a small L_k can be obtained by choosing a material with a low residual resistivity ρ_n and a high critical temperature T_c.[6] Geometry-wise, a short nanowire with a large cross-section wd is also beneficial for high count rates.

SNSPDs operating at GHz-count rates have already been demonstrated in 2004 [6]. Reset times of only 510 ps could be achieved by very short nanowires <1 μm [101]. The reset time can also be reduced by connecting multiple nanowires in parallel (for N nanowires in parallel, L_k is reduced by $1/N^2$ while maintaining the same covered area) [100, 102–104], enabling MCRs in GHz-range.

From Section 2.2.3, the reduction of L_k is limited by detector latching, which in turn is dependent on the hotspot relaxation time. Taking a typical relaxation time of 30 ps [54, 55], MCR is ultimately limited to ≈30 GHz. Apart from sufficiently long nanowires, latching can be prevented by shunt resistors in parallel to the detector. Recently demonstrated μm-wide detectors exhibit kinetic inductance values below 1 nH, thus require shunt resistors to avoid latching [49, 105]. Shunting the nanowire, however, has the disadvantage of lowering the amplitude of the output pulse, therefore reducing the signal-to-noise ratio.

When operating SNSPDs at high count rates, the gradual recovery of the current through the nanowire after a detection event has to be considered [99]. Consequently, the intrinsic detection efficiency does not reach its maximum immediately and the detection efficiency decreases above a certain count rate (see, e.g., Fig. 4d in Ref. [106]).

Another effect at high count rates is charge accumulation due to the capacitively coupled readout [107]. This leads to additional currents flowing through the nanowire, which reduce the measurable switching current [108].

[6] From Section 2.3.1, it is seen that these material properties are contrary to a large cut-off wavelength, thereby imposing a trade-off between detector speed and spectral sensitivity, at least when optimizing material parameters.

2.3.4 Other metrics

Timing jitter There is a latency between the photon impinging on the detector and the emergence of a voltage pulse. The probability density of this latency is called timing jitter and determines the timing resolution of the detector. The jitter value is usually given as the full width at half maximum (FWHM) of the probability density.

The overall system timing jitter comprises detector-related as well as system contributions (amplifier noise, fiber dispersion, instrumental factors). The detector-related contribution can be split into the longitudinal geometrical jitter (absorption site along the nanowire [109]) and the intrinsic jitter, which in turn consists of the transverse geometrical jitter (absorption site across the nanowire) and fluctuations within the detection process (see also summary in Tab. 1 of Ref. [110]). The intrinsic jitter and its dependence on the operating temperature, bias current, photon wavelength, and applied magnetic field have been studied extensively [110–115]. In 2020, a record timing jitter below 3 ps has been demonstrated with a 5 µm short nanowire, using free-space coupling and a cryogenic, low-noise amplifier [8].

Energy/Photon-number resolution Inherently, SNSPDs are not energy-resolving photon detectors like TESs. This can be traced back to the detection mechanism involving a large multiplication of quasiparticles. However, using a dispersive element in conjunction with an SNSPD, the spectral distribution of incoming light can be transformed into time information of the detected pulses to overcome this limitation. In Ref. [116], a graded-index optical fiber serves as the dispersive element. In this scheme, the temporal resolution (timing jitter) of the system directly translates to the spectral resolution. A single-photon spectrometer combining an on-chip grating with a long SNSPD has been shown by Cheng *et al.* [117]. The grating transforms spectral into spatial information which is then picked up by a long nanowire acting as a microwave delay line. Another approach for an SNSPD-based spectrometer was presented recently and does not require an additional dispersive element [18]. Instead, the nonlinearity of the detector over λ and I_b is used to compute the spectrum of the incident light.

The possibility of SNSPDs to resolve the photon number was first reported in 2017 by Cahall *et al.* [118]. Essentially, the photon number can be extracted from the pulse rise time of a single SNSPD, which is read out by a differentiating circuit.

Polarization sensitivity Due to the directional nature of meandered nanowires, SNSPDs are polarization-sensitive. Photons with the electrical field aligned along the nanowires are easier to detect than photons with a perpendicular orientation. This is mostly due to a higher absorption efficiency but an effect of the polarization on the intrinsic efficiency has been indicated as well [119]. To achieve the highest system detection efficiencies, the polarization is often tuned for maximum count rates [69].

Some SNSPDs have been specifically designed to increase the polarization sensitivity [120], with reported polarization extinction ratios above 400 [121]. The opposite approach, a detector with minimal polarization dependence, was adopted by designing spiral [122], orthogonally stacked [123], and fractal [124] nanowire geometries. A minimal polarization sensitivity has also been analyzed and designed by embedding the nanowire in a high-index dielectric material [125].

2.4 Detector materials

First demonstrations of SNSPDs have been shown using niobium nitride (NbN) as the detector material [4, 5]. NbN usually grows as a polycrystal, has a comparatively high critical temperature of (8 to 14) K for thin films and can therefore be operated at liquid helium (LHe) temperature of 4.2 K. However, its high T_c value goes along with a large energy gap $\Delta(0)$ of around 2 meV. Thus, NbN-based SNSPDs have limited sensitivity in infrared range, especially at high operating temperatures. For similar reasons, however, NbN has advantages when it comes to its timing characteristics. A record low timing jitter [8] and GHz-counting [6] have both been demonstrated using NbN-SNSPDs. In addition, its high switching current densities lead to high pulse amplitudes and therefore signal-to-noise ratios, which facilitates readout. It should be noted that, due to its early usage and ongoing research, NbN films have been optimized extensively for various detection metrics, resulting in largely varying material properties reported.

Mainly to further improve on the detector speed, niobium titanium nitride (NbTiN) has been investigated as an alternative material [126–129]. NbTiN has mostly comparable material properties to NbN, but exhibits a smaller kinetic inductance [126].

SNSPDs based on other nitrides like tantalum nitride (TaN) [96, 130–132] and titanium nitride (TiN) [133] have been reported as well. TaN is quite similar to NbN but has a lower critical temperature $T_c \propto \Delta$, thus a longer cut-off wavelength [132].

Besides nitrides, there have also been experiments using elemental superconductors for SNSPDs, in particular niobium (Nb) [134, 135]. Nb has a much smaller T_c of around 5 K for thin films [134], therefore also a smaller energy gap. However, its large diffusion coefficient of 1.4 cm^2/s (Tab. 11.4 in Ref. [51]), small residual resistivity $\rho_n < 100\,\mu\Omega\,cm$ [134, 135], and

Table 2.1: Parameter overview of materials commonly used for SNSPDs.

mat.	T_c (K)	ρ_n (μΩ cm)	$\lambda_l(0)$ (nm)	$\Delta(0)$ (meV)	D (cm²/s)	$C_e/C_{ph}\big\|_{T_c}$	ref.
Nb	4.5	6–80	161	0.7	1.4		[134, 135, 151]
NbN	7–14	100–600	250–400	2.0	0.45–0.7	0.57	[38, 66, 151]
NbTiN	7–14	100		2.0			[126, 152]
TaN	6–10.5	100–230	330–430	1.2	0.6		[132, 151]
WSi	3.7	238	700–800	0.6	0.58–0.75	5.65	[38, 72, 144, 153]
MoSi	3–6.5	160–240	950	0.4–0.9	0.48–0.67	5.21	[140, 141]
MgB_2	32	10–15	90				[147]

large electron density of states counteract its small energy gap and limit its energy sensitivity considerably. Moreover, Nb thin films require a protection layer to prevent oxidation.

Apart from nitrides and elemental superconductors, amorphous materials have been proposed to extend the energy sensitivity of SNSPDs. The lack of grain boundaries enables homogeneous thin films and can increase the fabrication yield, especially over large areas. Among the investigated materials, the main ones are tungsten silicide (WSi) [136, 137] and molybdenum silicide (MoSi) [138–141], but detectors based on NbSi [142] and MoGe [143] have been reported as well. In general, these amorphous materials have significantly lower critical temperatures of (2 to 6) K and saturated efficiencies well into mid-infrared range [11]. The enhanced sensitivity comes at the drawback of considerably worse timing characteristics, owing to the large kinetic inductance $L_k \propto \lambda_l^2$ ($\lambda_l \approx 750$ nm for WSi [144]) of these silicides. They also require working temperatures <2 K and a more elaborate readout because of the smaller output signal (smaller I_{sw}).

At last, SNSPDs out of materials with a much higher critical temperature are under research. Not only do they have the potential of requiring significantly less cooling power, but they might also have advantages in detection speed. The most promising candidate of these materials is magnesium diboride (MgB_2, $T_c = 32$ K), and SNSPDs based on MgB_2 show very small kinetic inductances, resulting in a reported reset time of only 130 ps [145–147]. The high-temperature superconductor $YBa_2Cu_3O_{7-x}$ (YBCO) is also actively researched as a material for SNSPDs [148, 149], but patterning YBCO nanowires poses a challenge [150]. More fundamentally, the large energy gap and much smaller hotspot sizes impedes photon detection at typical wavelengths and would necessitate much higher I_{sw}/I_{dep} ratios than conventional materials [38, 99].

In Tab. 2.1, the most important thin film parameters of common SNSPD materials are summarized.[7] Comparing the two primarily employed materials, NbN and WSi (representative of polycrystalline and amorphous materials, respectively), one can see the main contrast between detection speed and efficiency. While ρ_n and D are comparable between both materials, there are considerable differences in Δ, λ_l, and $C_e/C_{ph}\big|_{T_c}$. At a constant energy, an absorbed photon leads to a greater disturbance of the superconducting state in WSi compared to NbN. As described in Sections 2.2.1 and 2.2.2, this is attributed to a smaller energy gap Δ and a higher heat capacity ratio $C_e/C_{ph}\big|_{T_c}$ which leads to a shorter thermalization time but longer hotspot relaxation time in WSi. Consequently, NbN has a higher detection current I_{det} relative to I_{dep} [38]. On the other hand, the higher kinetic inductance and longer hotspot relaxation times of WSi reduce the ultimate limit of the maximum count rate [54].

There are different ways to tackle the problem of combining high detection efficiencies and good timing performance. One way to is to take the preferential timing characteristics of polycrystalline NbN and improve its detection efficiency by optimizing the relevant film properties. While reported near-unity intrinsic efficiencies up to 10 µm wavelength for WSi [11] seem out of reach for NbN, extending the spectral sensitivity up to few µm might be feasible. This can be achieved by uniform thin films with a small diffusion coefficient D, a moderate critical temperature T_c (balancing a high I_{sw} and small Δ), and especially a high current ratio I_{sw}/I_{dep}. The method and parameters used for the deposition of NbN thin films can have a large impact on its uniformity, stoichiometry, and crystal structure, all of which have an impact on the resulting material properties. Optimizing the deposition process is therefore often the most crucial step to improve the performance of SNSPDs.

2.4.1 State-of-the-art NbN thin film deposition

Reactive magnetron sputtering NbN is predominantly deposited by reactive magnetron sputtering. To deposit the film, a substrate is placed in a vacuum chamber, opposing the niobium target. Argon ions are accelerated by a direct voltage toward the target, which causes sputtering of the material. A magnet behind the target increases the ionization rate and enables lower working pressures. By adding nitrogen to the deposition chamber, nitrogen radicals can form a nitridized top layer on the Nb target which is then sputtered and deposited on the substrate. The relative content of Nb to N in the deposited film can be tuned by the partial pressure of nitrogen and the plasma discharge current [156].

[7] See also reviews of SNSPD materials in Refs. [99, 154, 155].

The substrate is commonly heated to temperatures in the range of (500 to 800) °C to increase the mobility of adatoms on the substrate surface [157]. In conjunction with lattice-matched substrates like sapphire (Al_2O_3) or buffer-layers like aluminum nitride (AlN), this can lead to mostly epitaxial film growth, accompanied by a lower sheet resistance and higher critical temperature. When the deposition temperature is lowered, the grain size gets smaller and the resistivity increases [158]. The process therefore relies on a homogeneous temperature distribution by the heater.

By careful optimization of the deposition parameters, the detection properties of NbN-based SNSPDs can be enhanced. It turned out that using Nb-rich films, higher I_{sw}/I_{dep} ratios up to ≈0.6 can be achieved, leading to a significantly lower dark-count rate and an increased cut-off wavelength [156]. Dane *et al.* showed a multi-step method to achieve smooth, polycrystalline NbN films and detectors with saturated count rates at 1550 nm [158]. Sputtering at room temperature instead of ≈800 °C lead, as expected, to critical temperatures <5 K and increased surface roughness. Applying an RF bias to the substrate holder compensated for these issues and produced smoother films with moderate T_c at ≈8 K for 5 nm thickness.

With sputtering however, scaling to larger wafers while retaining uniform film properties is difficult. The local deposition rate across the surface depends on the target size and the target–substrate distance. By increasing this distance, the area of uniform thickness can be increased. However, at the same time, the working pressure needs to be lower in order to sustain a ballistic deposition regime for which the mean free path of the sputtered particles is larger than the target–substrate distance. Reducing the working pressure calls for better vacuum conditions and is limited by the minimum pressure required for plasma ignition. Increasing the target size and the distance to the substrate also has practical limitations.

Atomic layer deposition As an alternative to sputtering, chemical vapor deposition (CVD) lends itself to deposit uniform films over large areas. Instead of a solid target material, volatile precursors are guided to the substrate and react with its surface. Among the CVD processes, atomic layer deposition (ALD) in particular is suited for the deposition of thin films and offers monolayer-control over the film thickness. In recent years, the interest in ALD as a deposition method for conformal thin films has grown steadily and has been adopted in various areas of nano- and optoelectronics, e.g., for high-k dielectrics (see overview in Refs. [159–161]).

Using ALD for the deposition of superconducting NbN films has been reported by several groups before. Originally, NbN was deposited by a *thermal* ALD process [162–167]. Thermal ALD requires rather high deposition temperatures and, for the deposition of NbN, uses a chlorine-based precursor ($NbCl_5$) as the niobium source [163]. Most of these studies used thick films ≥100 nm and suffered from contamination of chlorine in the films. Alén *et al.*

investigated thinner films of 10 nm and observed a decreasing chlorine content and resistivity when increasing the deposition temperature from 250 °C to 500 °C. Nevertheless, the use of chlorine-based precursors and reaction products complicates the deposition process and prevents the usage in various applications.

In 2010, Hinz *et al.* employed a plasma-enhanced ALD (PEALD) process using a metal-organic precursor for NbN films instead [168]. PEALD makes use of a plasma as one of the reactants, thereby enabling a higher reactivity and lower deposition temperatures compared to thermal ALD [161]. Depositing NbN by metal-organic PEALD got rid of chlorine in the process, however, the film quality suffered from high oxygen contamination.

In a similar PEALD process using a hydrogen plasma, Ziegler *et al.* were able to reduce the oxygen content in the films down to 11 at.% by reducing the hydrogen flow rate and optimizing the plasma exposure time [169]. The improved film composition was attended by a smaller room-temperature resistivity of 250 µΩ cm and a critical temperature above 10 K. This [169] and similar parameter studies by Sowa *et al.* [170], however, were still conducted on rather thick films with $d = (20 \text{ to } 40)$ nm.

Only recently, thinner ALD-NbN films <10 nm were studied [171, 172]. By further optimizing the plasma parameters (H_2 flow rate, exposure time), values of $T_c = 7.5$ K and $\rho(300\,\mathrm{K}) = 650$ µΩ cm have been measured on a 3.5 nm thick ALD-NbN on a sapphire substrate [171]. An increase in critical temperature (decrease in room-temperature resistivity) has been observed when increasing the film thickness until saturation above ≈15 nm. A strong dependence of the superconducting properties and the lattice parameter on the plasma exposure time has been observed as well [172].

Although high-quality thin films have been achieved, it remained unclear whether these films also serve the specific needs of SNSPDs. Specifically, crucial parameters like the diffusion coefficient D and current ratio I_{sw}/I_{dep} were not known and might differ substantially from the ones measured on sputtered samples. Moreover, the goal of using ALD-NbN for SNSPDs – large-area, uniform thin films to improve the efficiency and yield of detector arrays – needs to be verified.

Within this work, the suitability of ALD for the growth of NbN thin films has been investigated. Chapter 3 deals with the relevant film and microbridge parameters, including their spatial uniformity over the ALD sample table. In Chapter 4, these films are then used for the fabrication of SNSPDs and characterized optically. Beyond these basic characterizations, ALD-NbN SNSPDs are also used in experiments to test an optical coupling scheme (Chapter 6) and to demonstrate a method to extract current ratio I_{sw}/I_{dep} (Chapter 7).

2.5 SNSPD arrays

As specified in Section 2.3, many remarkable detection metrics have been shown using single-pixel SNSPDs. For applications in imaging, spectroscopy [173], or dark-matter search [19], however, *arrays* of SNSPDs are necessary.

Increasing the number of standard, individually biased SNSPDs is limited by the number of readout lines from the detectors at cryogenic temperatures to the bias and readout electronics at room temperature. There are both practical as well as thermal constraints to the number of lines. The heat load is limited by the cooling power of the cryogenic system and simply adding readout lines becomes unfeasible at some point.

Therefore, it is necessary to multiplex the detector operation when scaling up to hundreds or thousands of pixels. Because of bandwidth limitations, multiplexing comes with drawbacks to a number of performance metrics compared to single-pixel devices. Another distinction is the ability of photon-number or spatial resolution. Different multiplexing schemes have shortcomings in different metrics, which is why numerous schemes have been investigated.

The most advanced ones are briefly described in the next section, followed by a more detailed section on frequency-domain multiplexing, a technique that is employed and analyzed within this work.

2.5.1 Multiplexing schemes

Row-column multiplexing The most mature operation scheme for SNSPDs is row-column multiplexing [27, 174, 175]. In this scheme, each pixel consists of a nanowire together with a series resistor. The pixels are arranged in a grid and are connected in rows and columns. All pixels in a row are biased by a common current source. A detection event in a single pixel leads to a pair of voltage pulses of opposite polarities in the respective row and column. By matching pairs of pulses within a certain coincidence window, the event can be traced back to the triggering pixel [27]. The spatial information of a detection event is therefore encoded in the pulse arrival times on row and column lines. Compared to a direct readout of n pixels via n channels, this scheme only requires $2\sqrt{n}$ readout lines.

Row-column multiplexing for SNSPDs has first been demonstrated by a 2×2 proof-of-principle array in 2014 [174] and has later been expanded to arrays of 8×8 [175] and 32×32 pixels [27].

While a record number of SNSPDs has been read out using this scheme, there are also some drawbacks. Multi-photon events in different pixels can be registered but cannot be unambiguously assigned to the correct pixels. Apart from adapting the series resistance, pixels in a row can also not be biased individually and pixel-to-pixel variations due to fabrication imperfections can therefore not be compensated. Lastly, there is the issue of current redistribution after a detection event. This effect can increase timing jitter and lowers the signal-to-noise ratio, eventually limiting the number of pixels [175].

Building upon this multiplexing technique, Allmaras *et al.* demonstrated a thermally coupled row-column scheme which eliminates current redistribution and increases the fill factor [176].

Time-domain multiplexing Another multiplexing method utilizes the pulse arrival time to localize the triggering pixel in an array of serially connected detectors. This concept was first applied to SNSPDs by Hofherr *et al.* who showed a device which consists of two nanowires separated by a microwave delay line and is read out by a single readout line [177]. Based on the same idea, Zhu *et al.* showed an advanced, two-terminal device with 16 pixels and exploited the differences in pulse arrival times on both ends to resolve also two-photon events in different pixels [178]. Zhao *et al.* expanded this concept, utilizing the nanowire itself as a microwave delay line [63]. A maximum of 590 pixels over an active area of 286 µm × 193 µm has been resolved. The main drawback of time-domain multiplexing is the limited maximum count rate.

Pulse-amplitude multiplexing Amplitude multiplexing of SNSPDs can be accomplished by either connecting nanowires in parallel, each with a series resistor [100], or by pixels in series, each consisting of a nanowire and a parallel shunt resistor [179]. The number of simultaneously absorbed photons in different pixels can then by derived from the amplitude of the output signal. While a possible number of $\approx$100 pixels has been estimated [179], this method does not allow for spatial resolution.

SFQ In contrast to the multiplexing concepts above, which all rely on some form of analog multiplexing [26], there have also been demonstrations of on-chip digitization schemes by employing single-flux quantum (SFQ) circuits [180–182]. Although SFQ circuits add significant complexity to the readout and multi-photon events remain a challenge, promising results without detriment to the timing jitter have been shown [182].

2.5.2 Frequency-multiplexed SNSPDs

Frequency-division multiplexing (FDM) is a technique already established to read out arrays of superconducting detectors like TESs [183] or MKIDs [184, 185] (the latter featuring FDM intrinsically). Using FDM, an array of detectors can be read out by a single line, each pixel occupying a specific bandwidth. Thus, the spatial information of a detection event is encoded in the frequency domain. A similar method has been applied to SNSPDs by Doerner *et al.* [186].[8] In contrast to MKIDs, where the small change in inductance upon energy absorption causes a shift in the resonance, a frequency-multiplexed SNSPD (also called "radio-frequency SNSPD", RF-SNSPD) completely switches to the normal state, giving rise to a much larger signal amplitude. Therefore, the signal is easier to read out, however at the expense of energy resolution.

A maximum number of 16 pixels has been shown using this method, albeit only eight operating simultaneously [186]. Based on the inital work by Doerner *et al.* [28, 186, 188, 189], RF-SNSPDs are investigated further within this work in an effort to mitigate some of their shortcomings. The following sections deal with the basic concept of RF-SNSPDs and explain its pros and cons.

Operation principle and design An RF-SNSPD array consists of multiple pixels coupled to a common microwave transmission line. Each pixel represents a nanowire embedded in a resonator circuit which defines a resonance frequency unique to this pixel (Fig. 2.5a).

When a photon is absorbed, the nanowire switches to the resistive state and the resonant circuit's impedance increases significantly, equivalent to a strong decrease of the resonator's quality factor. This causes the energy saved in the resonant circuit to divert out and the signal to not couple into the resonator anymore, which can be registered on the readout line as a sudden increase in transmission S_{21} at f_r of the pixel. The result is an RF pulse with an envelope similar to a regular SNSPD pulse (Fig. 2.5b). Consequently, after current redirection out of the nanowire, the superconducting state gets restored and the probe tone returns back to the resonant circuit.

The time scale of the current restoration (and therefore the reset time) depends on the quality factor of the resonator. With a loaded quality factor of $Q = f_r/\Delta f$ and resonance width Δf

[8] As opposed to the frequency-multiplexing method described here, which uses the RF probe tone for both biasing and readout, there has also been shown an alternative method combining DC-biased SNSPDs with a microwave-multiplexed readout [187].

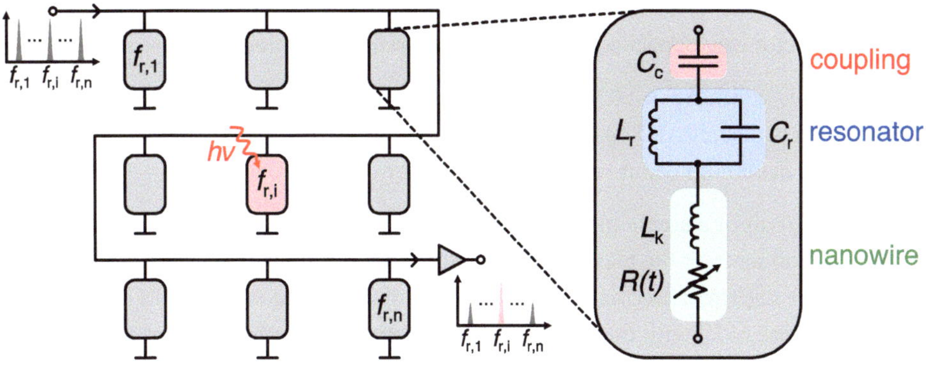

Figure 2.4: Schematic of a frequency-multiplexed SNSPD array. Each pixel consists of a nanowire (modeled by a time-dependent resistance $R(t)$ and kinetic inductance L_k) in series with a parallel LC resonant circuit (L_r and C_r which define a unique resonance frequency $f_{r,i}$) and a coupling capacitor C_c. All pixels are coupled to a common transmission line and are biased by an RF current consisting of tones of all pixels. A detection event in pixel i leads to an RF pulse with the pixel's characteristic frequency $f_{r,i}$.

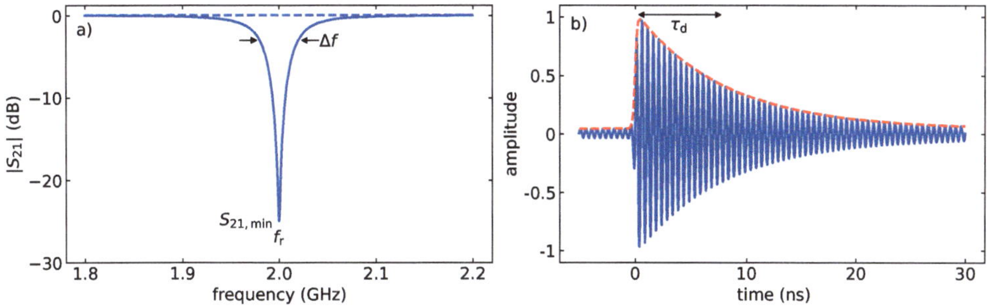

Figure 2.5: Simulated transmission and detector pulse of a single RF-SNSPD pixel. (a) Microwave transmission $|S_{21}|(f)$ on the transmission line with a resonance at $f_r = 2\,\text{GHz}$. Upon photon absorption, the resonance quickly vanishes (dashed line) which causes an RF pulse on the transmission line. (b) RF pulse (solid blue) and pulse envelope (dashed red) over time. The envelope qualitatively resembles the pulse shape of a regular SNSPD.

(taken at $-3\,\text{dB}$ of $|S_{21}|$), the decay time of the RF pulse (time of the current restoration to the resonant circuit) is given by

$$\tau_d = \frac{Q}{\pi f_r} = \frac{1}{\pi \Delta f}. \tag{2.13}$$

From Eq. 2.13, one can see the basic trade-off between the number of pixels fitting in a given bandwidth and the speed of an individual pixel. Pixels with a small quality factor Q have a short pulse decay time τ_d but occupy a large bandwidth Δf. On the other hand, pixels with a high quality factor can be packed more densely in frequency domain but are slower.

For a resonance width $\Delta f = 20\,\mathrm{MHz}$ and a rather conservative separation of 30 MHz, 100 pixels occupy a bandwidth of 5 GHz and each pixel has a dead time of ≈20 ns (MCR = 50 MHz). The combined maximum count rate for the whole array could therefore be ≈5 GHz.

The specific circuit of a single pixel shown in Fig. 2.4 was introduced by Doerner *et al.* to minimize the influence of the bias-dependent kinetic inductance $L_k(I_b)$ [188].[9] Originally, L_k of the nanowire itself was used as the inductive part of the *LC* resonant circuit [28]. However, due to an increase of L_k with the bias current, the resonance frequency $f_r = 1/\left(2\pi\sqrt{LC}\right)$ in this case shifted to smaller values. Because the current through the nanowire abruptly drops in case of a photon event, this can lead to a deficient detector operation when biasing at a constant frequency because the operation point cannot be reset. This problem is of particular importance if the resonance width is small, for example when designing larger arrays and therefore increasing the loaded quality factor $Q = f_r/\Delta f$ of each pixel. However, even with the optimized detector design, parallel inductance L_r (see Fig. 2.4) needs to be much larger than L_k in order to have a resonance frequency almost independent of L_k. Furthermore, the structure which defines L_r has to be designed much wider (here $w = 1\,\mu\mathrm{m}$) than the nanowire width. In this case, the bias power close to the switching current of the nanowire is still much lower than the one of the inductive structure, and L_r can be considered independent of the bias current. Hence, if the circuit elements are designed such that the shift in resonance frequency is much smaller than the resonance width Δf, dependence $L_k(I_b)$ does not impair the detector operation.

According to these considerations, the detector layout can be designed as follows:

1. Choosing the desired resonance width Δf and an array of resonance frequencies f_r.
2. In resonance, the imaginary part of the circuit impedance vanishes. Using the circuit of Fig. 2.4,

$$\mathrm{Im}(Z) = \omega_r L_k - \frac{1}{\omega_r C_c} - \frac{L_r/C_r}{\omega_r L_r - 1/\left(\omega_r C_r\right)} \overset{!}{=} 0. \tag{2.14}$$

   Assuming positive frequency values, there are two solutions to this equation. Here, only the smaller resonance f_r at which the capacitive impedance $1/\left(\omega_r C_c\right)$ dominates $\omega_r L_k$ is considered (see p. 73 f. in Ref. [190]).

[9] The dependence of L_k on current ratio I_b/I_{dep} will be utilized in Chapter 7 to extract I_{sw}/I_{dep}.

3. For Δf and f_r, values for C_c, L_r, C_r, and L_k are chosen according to Eq. 2.14. As mentioned above, L_r needs to be sufficiently large to ensure $L_r \gg L_k$. To set an array of resonance frequencies, either L_r or C_r can be varied.
4. Within this work, the resonant circuit consists of lumped elements. To acquire their physical dimension, the square kinetic inductance according to Eq. 2.4 is used for L_r and L_k, while C_c and C_r are defined by interdigitated capacitors with dimensions derived from previous simulations and measurements [188].

Pros and cons Apart from enabling the operation of multiple pixels via a single RF line, frequency-multiplexed SNSPDs have the advantage of an impedance matched to 50 Ω, which would require large tapers in the case of regular SNSPDs [191]. Furthermore, the bias level for each pixel can be individually set, which allows to compensate pixel-to-pixel differences due to nonideal fabrication. Compared to other multiplexing techniques, the fabrication is relatively simple because it does not require a multi-layer process. Lastly, multi-photon events can be registered and also correctly attributed to the triggering pixels.

Nevertheless, there are also several disadvantages of this multiplexing technique, requiring further development. The oscillating bias current leads to an oscillating operation point along the $\mathrm{CR}(I_b[t])$ curve and the intrinsic efficiency is therefore time-dependent. This was considered only recently [189] and its impact on the detection efficiency has to be tested. Moreover, switching from DC to RF bias implicates changes in the bias and readout circuit. Although bias tees to split the DC bias and RF signal are not required anymore and RF signals are easier to filter than DC signals, the generation of RF probe tones is much more complicated than a simple DC bias current. Single tones can be created by a signal generator, but operating an array of detectors requires a whole frequency comb. To create the bias signal for an RF-SNSPD array, an arbitrary waveform generator has been used before [186]. This might be facilitated by an FPGA (field-programmable gate array) together with a DAC (digital-to-analog converter), a method that has been used, e.g., for MKID arrays before [185, 192]. Also, extracting the spatial information, i.e., identifying the triggered pixel, is not trivial. This can be achieved by a Fourier transform of the detector signal and looking at the amplitudes in frequency domain. In Ref. [188], the detector signal was recorded and then transformed by a fast Fourier transform (FFT). At high count rates and in real-time, however, this method poses significant challenges for the readout system. Lastly, the resonant circuit of each pixels takes up a significant area on the chip and the fill factor is therefore very low.

This work Exploring the ramifications of the FDM scheme on the detector operation and possible solutions to the shortcomings of RF-SNSPDs will be the subject of Chapters 5 and 6.

The effect of the oscillating bias current on the detector operation is investigated in Section 5.2. The results are used to propose a gated detection scheme for synchronous applications, but also to estimate the impact on the detection efficiency in conventional free-running operation (Section 5.3). An improvement over the existing FFT method to extract spatial information by using correlation filters was suggested in Ref. [190] and is explored in Section 5.4. Finally, efficient optical coupling methods are introduced in the following section, and a method to improve the fill factor of RF-SNSPD pixels is presented in Chapter 6.

2.6 Optical coupling

Efficient photon coupling from the system input onto the active area of an SNSPD is difficult due to the commonly small active area of SNSPDs compared to other single-photon detectors. The coupling efficiency η_{oc} (see Eq. 2.9) is mostly dependent on the system around the detector instead of the detector itself, and the nanowire layout can only facilitate the coupling. In general, three different configurations can be distinguished: free-space, fiber, and on-chip coupling. In the following, the status quo of solutions for these cases are described.

2.6.1 Free-space coupling

To improve the efficiency of free-space coupling, longer nanowires covering a bigger area are the most straightforward method. This, however, goes along with a higher kinetic inductance and therefore smaller detection speed. Recently demonstrated micrometer-wide detectors from NbN [49] and MoSi [141] might solve this conflict because of much larger cross-sections and shorter lengths while covering a comparable area. Nevertheless, efficient free-space coupling remains a challenge which can, for the most part, be only solved by external optics and active alignment.

In Section 6.1, polymer microlenses directly on top of the active area are shown as a method to increase the effective collection area while keeping the nanowire length small. This is especially beneficial for frequency-multiplexed SNSPDs because their active area is small compared to the whole pixel size. The microlenses are fabricated by 3D direct laser writing (DLW), a technique that has been used to print free-form optical elements before [193].

2.6.2 Fiber coupling

Several methods have been shown to efficiently couple photons from optical fibers to SNSPDs. Micromachining the detector substrate to fit into a regular fiber sleeve is a method originally shown for the TES in 2011 [194] and has been widely adopted for SNSPDs since [69, 89, 139]. This method takes advantage of the high precision of photolithography (together with deep reactive-ion etching through a silicon substrate into a keyhole shape) and commercially available zirconia fiber alignment sleeves. Another coupling method using fibers adiabatically tapered down to a diameter of 1.3 μm and placed on top of a regular nanowire has been shown as well [195–197]. Utilizing the strong evanescent field, an overall SDE >50 % in a broad wavelength range (630 to 1500) nm has been achieved [197]. This, however, still requires active alignment and fixing of the microfiber, and may be impractical for an increased number of fibers.

There is also purely active alignment of fiber arrays with an SNSPD array of the same pitch. This has be achieved by observing the laser spot focused on the detectors by an infrared camera through the silicon substrate [198]. Other efficiency improvements of fiber-coupling include, e.g., band-pass filters integrated on the fiber end-faces [199] to suppress background radiation.

Lastly, there is also the possibility of fiber-to-waveguide coupling together with waveguide--integrated detectors (covered in the following section). Methods for fiber-to-waveguide coupling include grating couplers [71] and spot-size converters [200]. The latter has the advantage over traditional grating couplers of a much larger bandwidth.

Most of these methods, however, are relatively bulky or require active alignment and are therefore not well suited for arrays. The 3D-DLW technique, mentioned in the previous section for the fabrication of microlenses, can also be used to fabricate free-form optical waveguides for silicon photonics circuits, so-called photonic wire bonds (PWB) [201, 202]. Applying PWBs to SNSPDs would have the advantage of very compact packaging, mostly automatized fabrication without the need of precise active alignment between the components, and mode field adaption from the fiber to the active area of the detector. The possibility of using PWBs to couple fibers to SNSPDs is studied in Section 6.2.

2.6.3 On-chip coupling

Directly integrating SNSPDs with planar waveguides and using the evanescent field for photon absorption in the nanowire has the advantage of a much larger absorption length (several 10 μm) compared to normal incidence (≈5 nm). In addition to the above-mentioned fiber-based applications via fiber-to-waveguide coupling, waveguide-integration has also been shown in conjunction with an on-chip single-photon source [203]. Generally, on-chip integration has the advantage of scalability and minimizing lossy connections. However, a hybrid, multi-chip integration of SNSPDs with photonic circuits would be considerably more flexible, allowing for the combination of different materials and substrates, easier handling, and fewer constraints in detector fabrication (e.g., deposition temperature). Combining multiple chips, the above-mentioned PWBs have been demonstrated in the field of silicon photonic circuits for laser-to-chip [204, 205] and chip-to-chip [201] coupling. The integration of SNSPDs with photonic wire bonds presented for fiber-coupling in Section 6.2 might therefore be expanded for chip-to-detector coupling as well.

2.7 Summary

Superconducting nanowire single-photon detectors (SNSPDs) take advantage of a current-assisted breakdown of the superconducting state in a current-carrying nanostripe ("nanowire") to detect single photons from UV to mid-IR. They are distinguished from other types of single-photon detectors by their high detection efficiency, exceptional timing characteristic and low dark-count rates (see Tab. 2 in Ref. [206]). In this chapter, the basic detection mechanism and properties of SNSPDs were introduced. On the basis of the current status of theoretical models, the most important detector properties and their interaction were pointed out.

An efficient photon detection is favored by nanowires with a small, uniform cross-section wd, small electron diffusion coefficient D, large heat capacity ratio C_e/C_{ph}, low electron density of states $N(0)$, small energy gap Δ, and most importantly a high switching to depairing current ratio I_{sw}/I_{dep}. In general, smaller operating temperatures are also beneficial for the energy sensitivity of the detector. On the other hand, fast detectors with a small timing jitter can be achieved by short nanowires with a small kinetic inductance L_k. The latter corresponds to a big cross-section of the nanowire and a small penetration depth λ_l, which is realized by a small residual resistivity ρ_n and a large energy gap Δ.

It is seen from this comparison that there is a basic trade-off between detection efficiency and speed. By careful material choice, detector design, and fabrication optimization, the detector performance can be enhanced and adapted to the application. For individual, specifically designed detectors, record performances of near-unity system detection efficiency, GHz-count rates, and few-ps timing jitter have been demonstrated.

Scaling such single-pixel devices to detector arrays would open up the range of application of SNSPDs, in particular imaging. This requires further development in multiple areas: material and thin-film optimization, multiplexing techniques, and efficient optical coupling schemes. The state of the art in these three areas was summarized in the second part of this chapter to provide a basis of the following work.

Detector materials commonly used in SNSPDs and their different trade-offs were reviewed in Section 2.4. Polycrystalline and amorphous superconducting thin films were compared by their material properties relevant for SNSPDs. Niobium nitride (NbN) is the most widely used polycrystalline material and is suited for detectors with high timing performance but has drawbacks, particularly in detection efficiency, compared to, e.g., tungsten silicide (WSi). State-of-the art thin-film deposition of NbN was outlined and existing attempts to employ atomic layer deposition (ALD) as an alternative for the growth of homogeneous thin films were presented.

An overview of various demonstrated multiplexing schemes was given in Section 2.5. Each scheme enables the operation of multiple SNSPDs with a reduced number of readout lines but has a different set of shortcomings. The basics of the previously demonstrated frequency-domain multiplexing scheme for SNSPDs, so-called RF-SNSPDs, were explained. Instead of a DC-biased nanowire, RF-SNSPDs consist of nanowires embedded into resonant circuits and biased by a microwave current. This way, multiple detectors can be biased and read out by a single transmission line and spatial information on detection events is encoded in frequency-domain of the detector pulses.

Finally, existing schemes to efficiently couple photons from the detection system input to the commonly small active area of SNSPDs were introduced in Section 2.6. These schemes can be classified in free-space, fiber, and on-chip coupling. Most of the presented methods, however, have drawbacks in either scalability, size, bandwidth, or handling. Three-dimensional direct laser writing as an alternative fabrication technique for optical coupling elements and interconnects is already in use for silicon photonic circuits and its suitability for SNSPDs will be explored within this work.

3 Atomic-layer-deposited NbN thin films

Highly uniform superconducting thin films are beneficial for the fabrication of efficient single-pixel SNSPDs and necessary for arrays and large-area devices. Atomic layer deposition (ALD) is a deposition method with monolayer thickness control and a high degree of spatial uniformity. As outlined in Section 2.4.1, ALD-NbN thin films have been deposited and optimized before. In this chapter, the feasibility of ALD as a method to deposit NbN thin films is evaluated specifically for the needs and requirements to fabricate SNSPDs.

The first section describes the ALD process for NbN and the deposition parameters, which can have a significant influence on the properties of the growing films. Thereafter in Section 3.2, electrical properties of ALD-NbN films dependent on the film thickness and substrate material are investigated. Patterned micro- and nanostripes are characterized to assess the suitability of these films for SNSPDs. The influence of the plasma exposure on the film properties is investigated in Section 3.3 by varying the distance of the plasma source to the substrate during deposition. Finally, in Section 3.4, we analyze the spatial uniformity of ALD-NbN in terms of morphological, compositional, and superconducting properties to estimate its suitability for the fabrication of large-area, polycrystalline detector arrays.

3.1 Atomic layer deposition of NbN

The ALD process for NbN described here was developed by M. Ziegler and S. Linzen at Leibniz-IPHT, and the used deposition parameters are the result of previous optimizations [169, 171, 172].

Atomic layer deposition is based on the sequential exposure of the substrate to different gaseous reactants (precursors). These precursors react with sites on the substrate surface but can neither react with themselves nor with products of the surface reaction [207]. Purge steps between the exposure times prevent gas-phase reactions between subsequent reactants and ensure that the individual surface reactions within a single ALD cycle are separated in time. As with other CVD methods, ALD offers conformal coverage also for non-planar surfaces [161] but, in contrast to classic CVD, does not require precise control of the gas flow. Instead, the growth mechanism is self-limited by the available reactive sites on the substrate surface, leading to saturation after a specific time period. The only requirement is therefore the sufficient exposure of the reactant across the whole substrate surface [207]. Because of the self-limiting nature of ALD, there is a certain growth per cycle (GPC) and the film thickness can be set by the number of performed cycles, which results in a precise control of the film thickness at the atomic level. The ALD growth mechanism should, in theory, also enable uniform thin films over large areas.

One restriction to this ideal process is steric hindrance by the large functional groups of the precursor, which can block neighboring reactive sites on the substrate surface (i.e., the surface is saturated with adsorbed molecules before every reactive site is occupied). This leads to an average GPC below one monolayer. Other mechanisms causing a non-ideal film growth are reviewed in Ref. [208].

As outlined in Section 2.4.1, the most promising technology for the atomic layer deposition of NbN thin films is PEALD, which employs a plasma as the co-reactant. The specific PEALD process described here uses (*tert*-butylimido)-tris(diethylamino)-niobium (TBTDEN) as the first precursor which provides both the niobium as well as the nitrogen atoms for the growth of NbN. It consists of a niobium atom in the center surrounded by four ligands: three diethylamino groups ($-N(C_2H_5)_2$) and one *tert*-butylimido group ($=NC(CH_3)_3$). These ligands determine the properties of the precursor in a way that it is volatile at a moderate temperature and reacts with sites on the substrate but not with itself or the precursor layer already adsorbed to the surface [207]. The co-reactant is a hydrogen plasma generated by an inductively coupled plasma (ICP) source. The plasma constituents, hydrogen ions (protons) and radicals, react with the previously deposited precursor layer on the substrate surface. By forming volatile products (mainly C_xH_y) with the ligands of the adsorbed TBTDEN, they reduce the precursor layer and leave new reactive sites for the next ALD cycle.

There are several factors to consider for the deposition process. For the growth of NbN (or other nitrides), oxygen impurities can have a severe impact on the film properties. Oxygen incorporation in the films might be due to background oxygen in non-ideal vacuum conditions, oxygen impurities in the involved gases, or oxidation of the film after deposition (see Section III.E and Appendix B in Ref. [161]). The former contribution is exacerbated by the low deposition rate of ALD even for thin films (several hours for 5 nm), and by the high reactivity of niobium to niobium oxides and niobium oxynitrides. The latter contribution (ex-situ oxidation) could be caused, e.g., by formation of niobium hydrides at long plasma exposures, which then oxidize at the ambient air after taking out the sample. Another factor of particular importance is the time between the precursor and plasma exposure steps. At

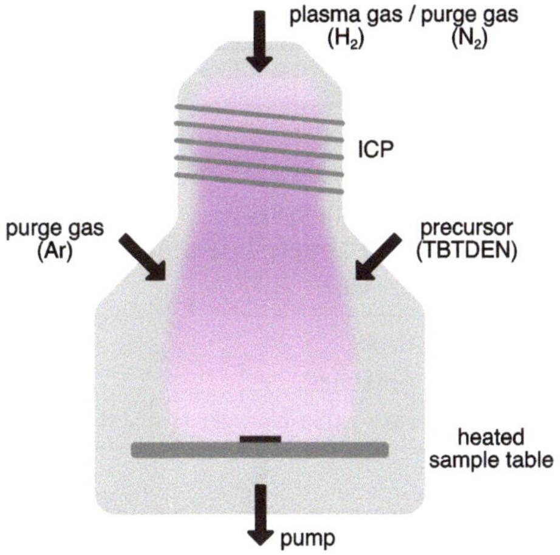

Figure 3.1: Sketch of the ALD chamber, consisting of a heated sample table, an inductively coupled plasma (ICP) source, and inlets for the various process and purge gases. The gases are purged through an outlet below the sample table.

the deposition temperature (here 380 °C), TBTDEN quickly decomposes pyrolytically into its organic compounds, forming carbides in the growing layer. It is therefore crucial to minimize the purge and pump time while still ensuring a clear separation between both exposures to maintain the ALD behavior.

In the following, the deposition process used within this chapter to fabricate ALD-NbN thin films is described. A sketch of the ALD chamber is shown in Fig. 3.1. All relevant parameters are summarized in Tab. 3.1.

For the deposition, the substrates are placed on the sample table (8-in. diameter) inside the ALD chamber, without the use of silver paste or other types of thermal adhesives. The chamber is evacuated to a base pressure of 2×10^{-3} mbar and the substrate table is heated up to 380 °C in about an hour. The precursor TBTDEN in the bubbler is pre-heated to 75 °C. The supply line from the bubbler to the ALD chamber is heated to a higher temperature of 120 °C to prevent condensation of the gaseous precursor. The actual ALD cycle starts with the TBTDEN exposure for a duration of 3 s. For this, argon is led through the bubbler containing the liquid precursor which is vaporized and gets – together with the inert argon gas – into the reaction chamber. A simultaneous nitrogen flow through the ICP on top of the chamber prevents the precursor from getting into the plasma source. This should have no impact on the surface reaction on the sample. In the first few cycles, hydroxyl groups (–OH) on the substrate surface act as reactive sites for the chemisorption of TBTDEN. In the subsequent

Table 3.1: Process parameters for the atomic layer deposition of NbN [169, 171, 172].

process step	parameter	value
preparation	sample table temperature	380 °C
	chamber wall temperature	150 °C
	bubbler temperature	75 °C
	precursor supply line temperature	120 °C
	base pressure	2×10^{-3} mbar
precursor	dose time	3 s
	total pressure	267×10^{-3} mbar
	flow rate ratio N_2 : Ar (bubbler)	1 : 3
1st purge	time	7 s
	total pressure	667×10^{-3} mbar
	flow rate ratio N_2 : Ar (purgc)	1 : 2
plasma exposure	time	60 s
	H_2 pressure	8×10^{-3} mbar
	plasma power	300 W
2nd purge	time	10 s
	total pressure	400×10^{-3} mbar
	flow rate ratio N_2 : Ar (p.) : Ar (b.)	5 : 10 : 1

cycles, amino groups ($-NH_2$) from previously deposited layers are the reactive sites. After the precursor exposure, excess precursor gas and reactants are purged for 7 s using argon and nitrogen. For the following plasma step, the hydrogen flow through the ICP inlet is set, after which the plasma is ignited with a power of 300 W and held for 60 s. The hydrogen plasma reacts with the organic ligands of the deposited precursor monolayer and reduces the deposited film to NbN. Following the plasma exposure, the volatile products of the surface reactions are purged with argon and nitrogen for 10 s and pumped out of the chamber. This concludes the ALD cycle, which is then repeated according to the desired film thickness.

3.2 Properties of ALD-NbN

This section contains parts from previously published content:
[209] E. Knehr, A. Kuzmin, D. Y. Vodolazov, et al., "Nanowire single-photon detectors made of atomic layer-deposited niobium nitride", Superconductor Science and Technology ***32****, 125007 (2019).*
Permission to include adapted text was obtained from the authors.
Data presented in Figs. 3.2 and 3.3 has also been published previously in the master's thesis of the author:
[210] E. Knehr, "Entwicklung Niobnitrid-basierter Einzelphotonendetektoren mittels plasmagestützter Atomlagenabscheidung", master's thesis (Karlsruhe Institute of Technology, 2018).
The ALD films were deposited by M. Ziegler and S. Goerke at Leibniz-IPHT. Sputtered films were deposited by A. Kuzmin at IMS. M. Ziegler patterned the bridges. E. Knehr measured and analyzed the samples with contributions from K. Ilin.

In Chapter 2, the influence of various material parameters on the detector performance in terms of photon sensitivity and detection speed was reviewed. For example, materials with a small superconducting energy gap $\Delta \propto T_c$, a small diffusion coefficient of the quasiparticles D, and a switching current I_{sw} close to the depairing current I_{dep} are advantageous to extend the SNSPD sensitivity into the infrared range. Material-inherent inhomogeneities and constrictions, on the other hand, can prevent an efficient photon detection and a high timing precision. Enabling highly uniform thin films with few-nanometer thickness and beneficial material properties is therefore the main goal of employing ALD to deposit NbN. Because of the entirely different growth mechanism of ALD compared to commonly used magnetron sputtering, the transport properties need to be investigated to assess the general suitability of ALD-NbN films for SNSPDs.[1] For this purpose, basic film and microbridge properties are measured in this section, depending on the film thickness, substrate material, and bridge width.

For sample preparation, the respective substrates were cleaned in an ultrasonic bath using acetone and isopropanol, before depositing NbN films of various thicknesses using the ALD process described in Section 3.1. Bare films prepared this way are characterized in Section 3.2.1. Additionally, micro- and nanobridges with four-point contacts were patterned for measurements on samples with defined dimensions in Section 3.2.2. The bridges were

[1] Previously, only ALD-TiN films have been reported for SNSPDs [133].

patterned by electron-beam lithography (EBL) using 130 nm thick negative-tone resist (ma-N 2401, micro resist technology GmbH). After development in AZ 726 MIF for 60 s, the resist pattern was transferred into the NbN layer by ion milling.

The films and micro-/nanobridges were characterized in a dipstick equipped with DC lines and cooled down to 4.2 K in LHe. For magnetic field measurements, a pulse-tube cryocooler equipped with a superconducting magnet was used instead. The coils create a magnetic field up to 5 T perpendicular to the substrate surface of the samples.

3.2.1 Film properties

The basic electrical characteristics were investigated on unpatterned ALD-NbN films with a thickness of (3 to 21) nm, deposited on sapphire and quartz substrates with a size $w \times l =$ 2 mm × 10 mm.

In Figs. 3.2a and 3.2b, the sheet resistance R_{sh} in the normal state near the superconducting transition and the residual resistivity $\rho_{\mathrm{n}} = R_{\mathrm{sh}}d$ are plotted as a function of thickness d. Both are only weakly dependent on the film thickness for $d \geq 5$ nm. For thinner films, however, both parameters rapidly increase. The same qualitative dependence is observed for sapphire and quartz substrates, with films on quartz exhibiting slightly higher values of R_{sh} and ρ_{n}. On sapphire, the lowest value of ρ_{n} of about 300 µΩ cm corresponds to a film thickness of 6 nm.

In Figs. 3.2c and 3.2d, the critical temperature T_{c} (taken at $0.01R_{\mathrm{sh}}$) and the width of the transition ΔT (temperature difference of $0.9R_{\mathrm{sh}}$ and $0.1R_{\mathrm{sh}}$) from the superconducting to the normal state are plotted over thickness d. Similar to the trend of R_{sh}, the critical temperature T_{c} reaches a plateau above $d \approx 5$ nm. A value of $T_{\mathrm{c}} = 11.3$ K for a 6 nm thick film has been obtained, while the critical temperature decreases for smaller thicknesses, reaching 7.4 K at 3 nm. The rapidly degrading properties toward thinner films can be explained by the shrinking size of the superconducting core between the two normal-conducting layers at the surface and near the substrate interface, which are assumed to be constant in size for each NbN thickness (proximity effect, see Refs. [66, 211, 212]). Again, the values for films on quartz are slightly worse, having a ≈1 K smaller T_{c} and slightly larger ΔT for films with $d < 10$ nm. This difference decreases for thicker films, for which the substrate interface plays a decreasing role regarding the film properties.

With ρ_{n} and T_{c}, the London penetration depth λ_{l} can be estimated by (Eq. 2.5)

$$\mu_0 \lambda_{\mathrm{l}}^2 = \frac{\hbar \rho_{\mathrm{n}}}{\pi \beta_0 k_{\mathrm{B}} T_{\mathrm{c}}} \tag{3.1}$$

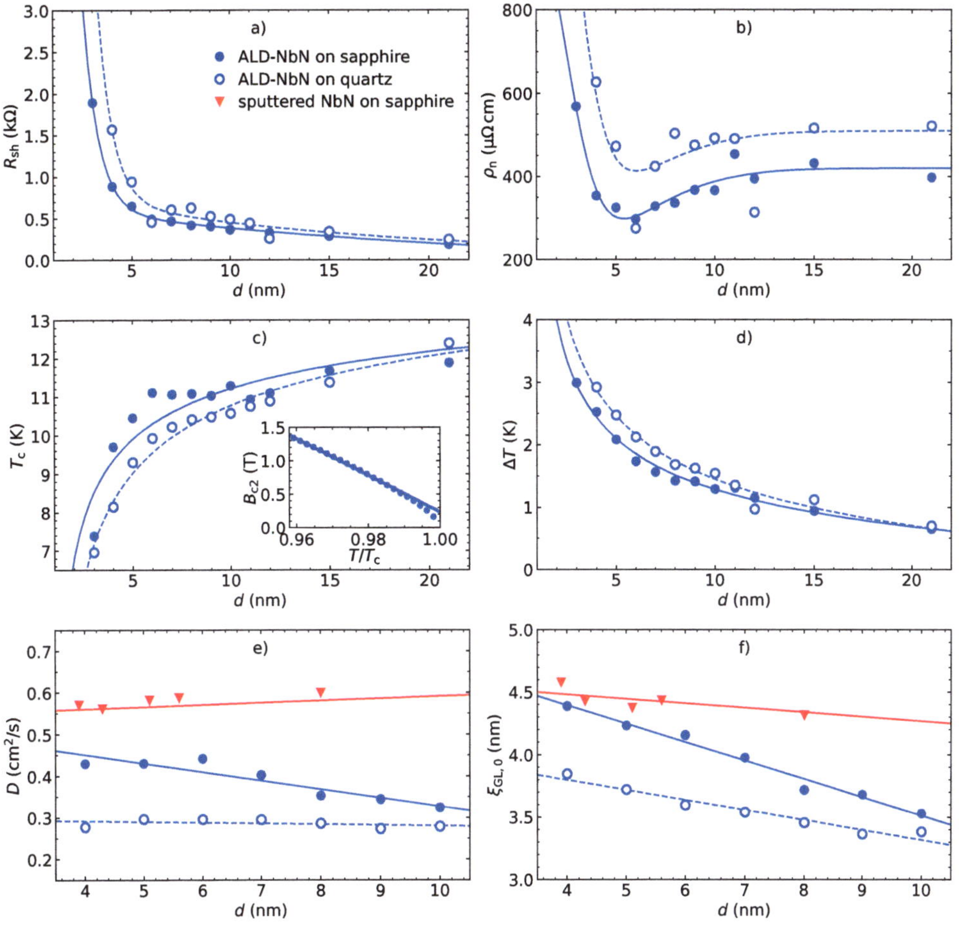

Figure 3.2: Electrical properties of ALD-NbN films with varying thickness on sapphire and quartz substrates: sheet resistance R_{sh} in normal state (a), residual resistivity ρ_n (b), critical temperature T_c (c), transition width ΔT (d), electron diffusion coefficient D (e), and the Ginzburg-Landau coherence length at zero temperature $\xi_{GL,0}$ (f). The solid and dashed lines are guides to the eye. Values from sputtered NbN in (e) and (f) are taken from Ref. [66]. [Inset in (c)] Exemplary temperature dependence of the second critical magnetic field B_{c2} of the 5 nm thick NbN sample on sapphire; the solid line represents a linear fit excluding data points above $0.99\,T/T_c$.

and amounts to ≈550 nm for 5 nm and 10 nm thick films on sapphire ($\beta_0 = 2.05$).

To obtain the electron diffusion coefficient D and the Ginzburg-Landau coherence length ξ_{GL}, the second critical magnetic field B_{c2} was measured dependent on the temperature. For this purpose, an external magnetic field normal to the sample surface was applied. Here, T_c was defined at $0.5R_{sh}$, while B_{c2} was taken as the magnetic field at which the same resistance

$0.5R_{sh}$ was reached for $T < T_c$. Exemplary experimental data on $B_{c2}(T)$ of the 5 nm thick film on sapphire is shown in the inset of Fig. 3.2c. In the observed temperature range, $B_{c2}(T)$ is mostly linear. Only very close to T_c ($T \geq 0.99T_c$), the measured values deviate from this dependence, an effect attributed to the deformation of the vortex lattice due to the diverging coherence length and the granular film structure [213]. At temperatures close to 0 K, on the other hand, the second critical magnetic field is expected to saturate [214]. Instead of extrapolating the linear dependence to 0 K, $B_{c2}(0)$ was therefore obtained by

$$B_{c2}(0) = 0.69\, T_c \left.\frac{\mathrm{d}B_{c2}}{\mathrm{d}T}\right|_{T \lesssim T_c}, \tag{3.2}$$

see Fig. 1 (dirty limit) in Ref. [214].

The temperature dependence $B_{c2}(T)$ was acquired for the film thickness range (4 to 10) nm. From slope $\mathrm{d}B_{c2}/\mathrm{d}T$, coefficient D is then calculated by

$$D = -\frac{4k_B}{\pi e}\left(\frac{\mathrm{d}B_{c2}}{\mathrm{d}T}\right)^{-1}, \tag{3.3}$$

see appendix in Ref. [66]. With $B_{c2}(0)$ from Eq. 3.2, the Ginzburg-Landau coherence length at 0 K is estimated by

$$\xi_{GL}(0) = \sqrt{\frac{\Phi_0}{2\pi B_{c2}(0)}}, \tag{3.4}$$

with the magnetic flux quantum $\Phi_0 = h/2e$ (see Eq. A6 in Ref. [66]).

Figure 3.2e shows the diffusion coefficient D dependent on the film thickness. On both sapphire and quartz substrates, the obtained values for ALD-NbN are smaller than reference values from sputtered NbN (taken from Tab. 1 in Ref. [66]). Comparing the values of films on sapphire, D of ALD-NbN is about 25 % smaller at 5 nm and almost 50 % smaller at 10 nm. Contrary to the expectation that the diffusion coefficient increases with film thickness (as the crystalline structure can be assumed to improve), the extracted values D actually slightly decrease for thicker films on sapphire. One reason could be oxygen incorporation into the films as a result of the long ALD deposition times together with oxygen permeation into the deposition chamber. This causes growing defects (niobium oxide) in the deposited film and could, thus, lead to a decrease of the diffusion coefficient. This thickness dependence, however, was not observed on quartz, for which D is almost constant at $0.3\,\mathrm{cm^2/s}$. As described in Section 2.2.1, a smaller diffusion coefficient is beneficial for photon detection because it confines the absorbed photon energy in a smaller volume and, subsequently, leads

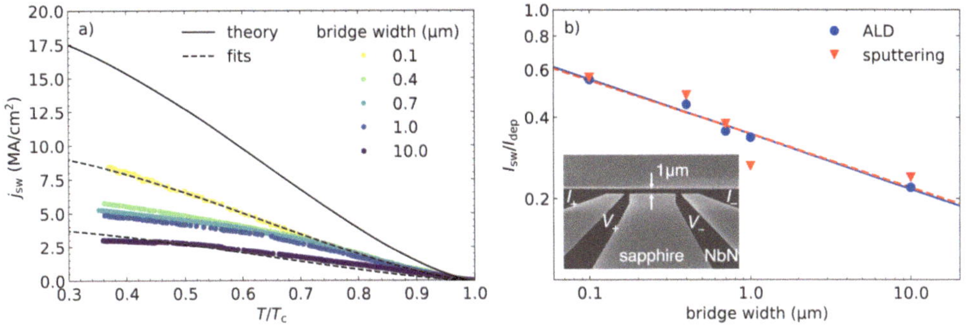

Figure 3.3: (a) Temperature dependence of switching current density j_{sw} for different bridge widths ($d = 10$ nm). The solid line represents $j_{dep}(T)$ according to Eq. 3.5, while the dashed lines are fits to the data for $w = 0.1\,\mu m$ and $w = 10.0\,\mu m$ with $j_{dep,GL}(0)$ in Eq. 3.5 as fitting parameter. (b) Dependence of the relative switching current (ratio of measured I_{sw} to theoretical I_{dep}) on bridge width of ALD and sputtered NbN ($d = 10$ nm). The solid and dashed lines are to guide the eye. (Inset) Exemplary SEM image of the bridge geometry for four-point measurements (here $w = 1.0\,\mu m$).

to a stronger suppression of the order parameter in the hotspot and a shorter thermalization time [38].

Compared to reported values for sputtered films [66], the diffusion coefficient is 1.3 to 2.0 times smaller, while the residual resistivity of ALD-NbN films is about two times higher for the same film thickness. Consequently, this should be attended by a similar single-spin density of states at the Fermi level $N(0) = 1/(2e^2 \rho_n D)$. As a result, the heat capacity of the electronic system $C_e \propto N(0)$ and, thus, the heat capacity ratio C_e/C_{ph} should be similar to sputtered films [38]. As described in Section 2.2.1, a larger ratio C_e/C_{ph} facilitates photon detection because of the larger energy fraction transferred to the electronic system.

In Fig. 3.2f, the coherence length ξ_{GL} is plotted, and a decrease for thicker films is observed. This is expected due to the increase in T_c and the relation $\xi_{GL}(0) \propto \sqrt{\xi_0 l_0}$, where $\xi_0 \propto T_c^{-1}$ (ξ_0 being the BCS coherence length and l_0 the mean free path of electrons; see appendix in Ref. [66]). The value $\xi_{GL}(0) = 4.23$ nm at $d = 5$ nm of ALD-NbN coincides well with a reference value ($\xi_{GL}(0) \approx 4.37$ nm for $d = 5.1$ nm) of sputtered NbN films of the same thickness on sapphire substrates [66]. On quartz, the ALD-NbN films exhibit slightly smaller values of $\xi_{GL}(0)$ and a weaker dependence on the film thickness.

3.2.2 Micro- and nanobridge properties

From the NbN films characterized in the previous section, samples with 10 nm thick NbN on sapphire were chosen to pattern bridges for four-probe measurements. The bridges have a width w between 0.1 µm and 10 µm and a length $l = 10w$ (10 squares, see the geometry in the inset of Fig. 3.3b). A reduction of the measurable switching current I_{sw} due to a nonuniform current distribution across the bridge is neglected since $w \ll \Lambda \approx 60$ µm (with Pearl length $\Lambda = (2\lambda_1^2)/d$ for thin films with $d \ll \lambda_1$) [215, 216]. For comparison, bridges of the same geometry were fabricated from sputtered NbN and etched in parallel to the ALD-NbN samples. The magnetron sputtering process is described in Ref. [156].

In Fig. 3.3a, the dependence of j_{sw} on the relative temperature T/T_c is shown. Near T_c, the $j_{sw}(T)$ curves for all widths coincide as expected from the Ginzburg-Landau theory. At lower temperatures, the penetration and dissipative movement of vortices across the bridge suppress the measurable value of j_{sw} for wider bridges because of the lower energy barrier for vortex penetration [217]. Therefore, the curves diverge toward smaller T/T_c.

The experimental values are compared to the temperature dependence of the depairing current density $j_{dep}(T)$ according to the model of Kupriyanov and Lukichev (KL) [218]

$$j_{dep}(T) = j_{dep,GL}(0)\left(1 - \frac{T}{T_c}\right)^{\frac{3}{2}} \mathrm{KL}(T), \tag{3.5}$$

with the formal Ginzburg-Landau (GL) zero-temperature depairing current density

$$j_{dep,GL}(0) \approx 1.454\, \beta_0^2 \frac{(k_B T_c)^{\frac{3}{2}}}{e\rho_n\sqrt{D\hbar}} \tag{3.6}$$

and the correction factor

$$\mathrm{KL}(T) = 1.87 - \frac{1.46}{1 + (0.7\frac{T}{T_c})^{1.12}}. \tag{3.7}$$

The correction factor KL(T) is an approximation of curve 7 (dirty limit) in Fig. 1 of Ref. [218].

The theoretical $j_{dep}(T)$ curve calculated by Eqs. 3.5, 3.6 and 3.7 (with experimentally obtained values for T_c, ρ_n, and D) is shown as a solid line in Fig. 3.3a. In addition, fits to the experimental data for $w = 0.1$ µm and $w = 10.0$ µm are also plotted (dashed lines). Here, parameter $j_{dep,GL}(0)$ is used as a fitting parameter. As it is seen, the data for $w = 0.1$ µm can be well fit by Eq. 3.5, which could indicate that ratio j_{sw}/j_{dep} is only reduced by constrictions. In this case, the fit estimates the cross-section of the effective superconducting core of the bridge and j_{sw} coincides with the critical current density j_c. For the widest structure (10.0 µm), however, the fit does not resemble the shape of the experimental curve, which is most likely limited by vortices at lower temperatures.

In Fig. 3.3b, the ratio I_{sw}/I_{dep} at 4.2 K is plotted over the stripe width for both ALD and sputtered NbN bridges. The samples of both deposition technologies exhibit almost coinciding dependences on the width, with I_{sw}/I_{dep} ratios growing from ≈0.2 for the widest samples up to almost 0.6 for the narrowest ones. Although the absolute values of j_{sw} of the ALD-NbN bridges were significantly lower than the ones for sputtered NbN (9 MA/cm^2 compared to 23 MA/cm^2 for $w = 0.1$ µm), due to a higher ρ_n and lower T_c (see Eq. 3.6), the calculated ratios I_{sw}/I_{dep} are comparable, reaching 0.55 and 0.56 at the minimal width of 0.1 µm.

The obtained ratios are still below one even for $w = 0.1$ µm, which is expected and probably due to both material-inherent nonuniformities and fabrication imperfections of the nanowire geometry. For the cross-section, we use the nominal width determined by scanning electron microscopy (SEM) without taking the edge roughness into account. Also, damaged normal-conducting bands around the superconducting core were neglected. These are expected due to film oxidation, the formation of an intermediate layer between the substrate and the film, and damaged edges during etching, according to the proximity effect described in Ref. [211]. Ratio I_{sw}/I_{dep} can, thus, be treated as key figure to quantify the material and geometrical constrictions, i.e., the uniformity of the superconducting structure.

3.3 Variation of the ICP-to-sample distance

The NbN films were deposited by M. Ziegler and V. Ripka at Leibniz-IPHT. Four-point bridges were patterned by M. Ziegler (EBL) and E. Knehr (etching). P. Schanz and E. Knehr measured and analyzed the samples.

From previous investigations on the influence of various ALD parameters on the film properties [169, 171], the strong dependence of the NbN film quality on the hydrogen-plasma condition (hydrogen flow rate, exposure time) has become evident. In this section, the plasma influence is further investigated by varying the vertical distance between the remote ICP source and the substrate (Fig. 3.1), thereby modifying how the plasma affects the film growth, while keeping the other deposition parameters the same (see the caveats to this in the discussion below). By bringing the sample closer to the ICP, the interaction of the plasma species (ions, radicals) with the precursor layer on the substrate surface should be amplified. This method was also chosen because in the current ALD process the ICP is already set to the highest possible power (300 W) of the particular source.

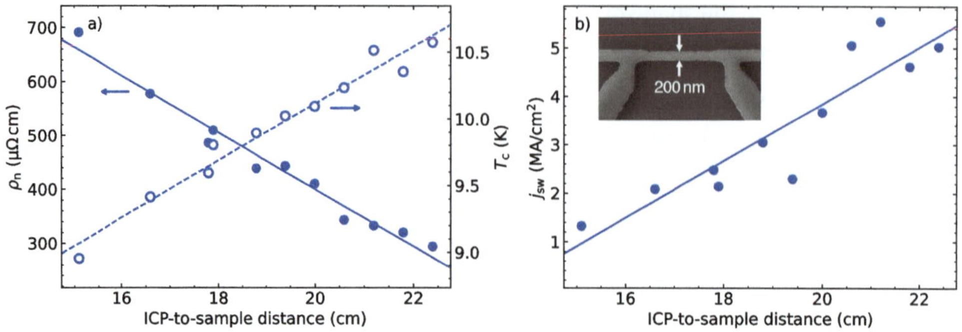

Figure 3.4: Properties of ALD-NbN stripes dependent on the distance between the inductively coupled plasma (ICP) source and the substrate: residual resistivity ρ_n, critical temperature T_c (a), and switching current density j_{sw} (b). Each NbN stripe is 10 nm thick, 200 nm wide and 2 μm long [see inset in (b)].

The vertical distance of the ICP from the substrate table amounts to 23 cm. By placing the samples on aluminum pedestals of various heights (between 0.6 cm and 7.9 cm), this distance was reduced. Each pedestal was placed in the center of the ALD sample table concentrically below the ICP, and each sample was deposited in a separate run. As substrates, we used 10 mm × 10 mm silicon chips with 300 nm thermal oxide and a 20 nm thick Al_2O_3 layer deposited by ALD beforehand and in parallel for all substrates. Each NbN layer had a nominal thickness of 10 nm.

After deposition, nanobridges for four-point measurements were patterned from each film by EBL and ion milling (same process as described in Section 3.2). The following measurements were conducted on bridges with a size $w \times l = 0.2\,\mu m \times 2\,\mu m$. Similar to Section 3.2, standard $R(T)$ and $I(V)$ measurements were conducted to extract the basic superconducting properties of the samples.

The results are depicted in Fig. 3.4. As it is clearly visible, there is a linear dependence of all parameters on the distance of the sample to the ICP source. The "best" properties (lowest ρ_n, highest T_c and j_{sw}) were obtained on the sample the furthest away from the ICP source. Extrapolating the measured dependence, these values are consistent with previously measured ones from samples placed directly on the same table at distance 23 cm but on sapphire and quartz substrates (compare to Figs. 3.2 and 3.3). The closer the substrate was placed to the plasma, the stronger the degradation of the superconducting properties. At the closest distance of 15.1 cm, the residual resistivity has increased up to 690 μΩ cm (up from 293 μΩ cm), while T_c has decreased to 8.95 K (down from 10.57 K) and j_{sw} to 1.32 MA/cm^2 (down from $\approx$5.5 MA/cm^2).

One reason that could explain the observed degradation is the higher plasma power which might lead to hydrogenation of the deposited film (NbH_n). These hydrides can oxidize to compounds like Nb_2O_5, potentially leading to a significant amount of oxygen impurities and a degradation of the NbN film properties. This assumption could be substantiated by studying the compositional changes of the films and, specifically, a possible increase in oxygen content in films on substrates placed closer to the ICP.

However, there are other factors beyond the immediate plasma influence in this particular parameter-variation experiment which could explain the observed film degradation. The substrate table is heated to 380 °C. The aluminum pedestals which are used to vary the height, however, likely cause the substrate temperature to be significantly lower. A lower deposition temperature was found to have a strong influence on the film properties before [169]. On the other side, there is a stronger heat contribution on the substrate surface by the plasma itself for samples closer to the ICP, which might have an influence on the film growth as well. Another factor could be a different precursor-gas exposure of the substrates at various heights, which is, however, highly unlikely given that the ALD process is mostly independent of the specific gas distribution.

The influence of the substrate's proximity to the remote plasma is therefore not yet conclusively evaluated and further investigations are necessary to isolate the various possible contributions to the film growth. Based on these results and considering that the distance is limited by the physical dimensions of the ALD chamber, lowering the plasma power might be worth to consider for future process optimization.

3.4 Spatial uniformity of ALD-NbN

*This section contains parts from previously published content:
[219] E. Knehr, M. Ziegler, S. Linzen, et al., "Wafer-level uniformity of atomic-layer-deposited niobium nitride thin films for quantum devices", Journal of Vacuum Science & Technology A **39**, 052401 (2021).
Permission to include adapted text was obtained from the authors. The experimental concept was developed by E. Knehr together with S. Linzen. The films were deposited by M. Ziegler and V. Ripka, bridges were patterned by M. Ziegler (EBL) and E. Knehr (etching). The measurements were performed by M. Ziegler (AFM), U. Brückner, J. Plentz (SIMS), S. Linzen, M. Diegel (XRD), P. Schanz, and E. Knehr (electrical). The experimental results were analyzed by E. Knehr with contributions from M. Ziegler, S. Linzen, J. Plentz, and K. Ilin.*

The film uniformity is crucial for the fabrication of SNSPDs. Nanowire constrictions limit the photon sensitivity of the detector as they reduce the maximum current the nanowire can carry, thus, only a fraction of the nanowire is optimally biased. Apart from the effect on the minimal detectable photon energy, there's also a detrimental impact on the detector's timing jitter [114, 220, 221]. These constrictions are not only caused by patterning imperfections, i.e., edge roughness, but also small variations in thickness and film-intrinsic fluctuations in composition and parameters like the superconducting gap [75, 222, 223].

Another factor with similar consequences on the detector performance is the large-area film uniformity. This becomes important when scaling SNSPDs to larger areas and pixel numbers. A degradation in film properties at a distance from the optimum area (e.g., the center of the wafer table in a deposition chamber) can be compensated for only by few multiplexing architectures and also only to a certain degree.

As discussed before, ALD lends itself well to deposit smooth and uniform thin films over large areas due to its self-limiting, surface-controlled growth mechanism [207]. Nevertheless, there are limitations, e.g., imposed by the plasma distribution of the PEALD process used for NbN. For this reason, the parameter spread of NbN thin films regarding their morphology, composition and superconducting properties over the equivalent area of a 6-in. wafer is studied in this section.

The same deposition parameters as specified in Tab. 3.1 are employed. As substrates, we use (100) silicon with 300 nm thermal oxide. A total of 25 silicon chips of 10 mm × 10 mm size were arranged in a cross-like shape with the remote plasma source (75 mm diameter)

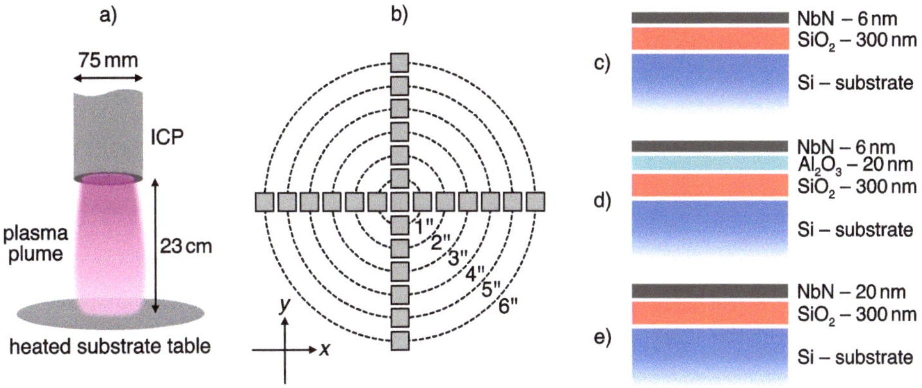

Figure 3.5: (a) Dimensions of the ICP configuration in the ALD chamber (not to scale). (b) Chip distribution across the ALD wafer table. Each grey square represents a 10 mm × 10 mm chip. (c), (d), (e) Film-layer stacks used to investigate the spread in electrical parameters and surface roughness (c), in film thickness (d), and in composition and crystal structure (e); film thicknesses not to scale.

concentrically placed 23 cm above the center. In the following, the x- and y-coordinates correspond to this position during deposition, $x, y = (0, 0)$ being defined as the center of the substrate table. Three depositions were conducted: one with a target thickness of 6 nm (130 ALD-cycles) for atomic force microscopy (AFM) and electrical measurements, another 6 nm deposition with a 20 nm thick etch-stop Al_2O_3 layer deposited prior to NbN for thickness measurements, and one with a thickness of 20 nm (435 ALD-cycles) for x-ray diffraction (XRD) and secondary ion mass spectroscopy (SIMS) measurements. The different film-layer stacks of each deposition together with schematics of the sample arrangements are shown in Fig. 3.5.

Electrical four-point probe measurements were conducted on microbridges patterned from the NbN films to determine temperature-dependent resistivity and switching current. The NbN bridges are 1 µm wide and 10 µm long (see geometry in the inset of Fig. 3.8). This comparatively wide structure was chosen to separate film-inherent inhomogeneities from any adverse effect caused by the patterning process. To avoid current crowding and maximize the switching current [82], the voltage leads to the microbridge are rounded off. The patterns were defined by electron-beam lithography using 130 nm thick ma-N 2401 resist exposed to electrons of acceleration voltage 50 kV with a dose density of 495 $\mu C/cm^2$. After development, the written structures were transferred into the NbN layer by Ar-ion milling.

The NbN film thickness was measured on steps etched into the samples with the additional Al_2O_3 layer by reactive-ion etching (RIE) using CF_4 and oxygen. With an etch rate of

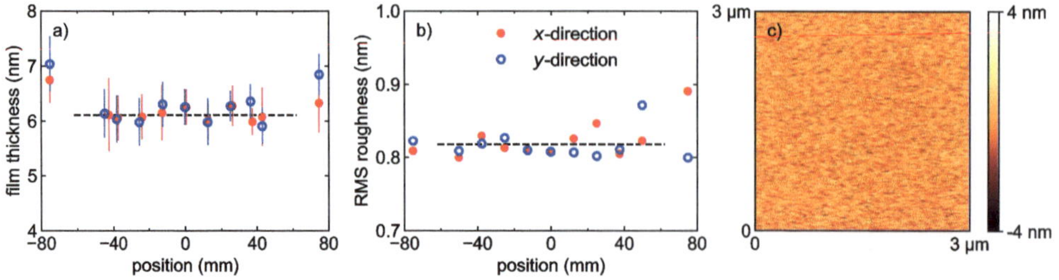

Figure 3.6: (a) Spread of film thickness over the position across the wafer table. The error bars correspond to 95 % confidence intervals. The horizontal dashed line represents the mean value of 6.1 nm within ±50 mm from (0, 0). (b) Distribution of root mean square (RMS) surface roughness over a 6-in. area. The horizontal dashed line represents the mean value of 0.82 nm within ±50 mm. (c) Exemplary 3 µm × 3 µm AFM image at position $y = -37.5$ mm with the mean value as zero level.

0.05 nm/min compared to 18 nm/min of NbN for etching conditions at hand, the Al_2O_3 layer acts as an etch-stop layer and therefore allows for a well-defined step.

3.4.1 Morphological uniformity

To evaluate geometric film homogeneity, we measured the film thickness and surface roughness using an FRT MicroProf tool equipped with an atomic force microscope. The surface was scanned in the non-contact mode with a line speed of 9 µm/s. The film thickness was defined as the mean value of 21 individually measured profiles of the patterned steps on each chip. The surface roughness was taken as the root mean square (RMS) roughness over a scanned area of 5 µm × 5 µm at various positions.

Figure 3.6a shows the distribution of film thickness along the x- and y-axis of the examined area. Within a diameter of 4 in. ($|x, y| \leq 50$ mm), there is a negligible variation in thickness around a mean value of 6.1 nm, equivalent to a film inhomogeneity of 3.6 %. This corresponds to a growth rate of 0.469 Å/cycle, which is comparable to previously reported values [170, 171]. Increasing the area to 6 in. leads to 9.1 % inhomogeneity and a maximum thickness of (7.0 ± 0.3) nm ($y = -75$ mm).

The NbN surface roughness is shown in Fig. 3.6b, with an exemplary AFM image in Fig. 3.6c. The RMS roughness R_q fluctuates around a mean of 0.82 nm, with a maximum of 0.89 nm at $x = +75$ mm. R_q corresponds to roughly two lattice constants of NbN (0.45 nm [224]). A reference measurement of the bare substrate led to similar $R_q = 0.8$ nm. This indicates that ALD-NbN roughness is mainly determined by the surface roughness of the substrate below.

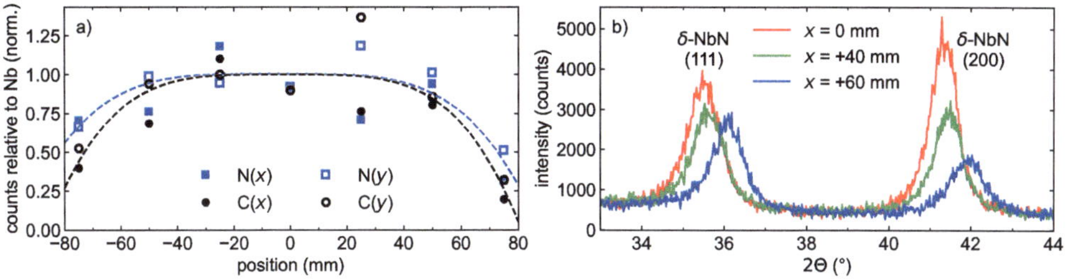

Figure 3.7: (a) SIMS: distribution of nitrogen and carbon counts relative to the respective niobium count at each position. Both relative curves are normalized, i.e., the absolute values of nitrogen and carbon cannot be compared. The dashed lines are guides to the eye. (b) XRD: 2Θ-scan (ω fixed at $0.4°$) of 20 nm thick ALD-NbN at three positions along the x-axis of the wafer table. The two peaks around $35.5°$ and $41.5°$ correspond to orientations (111) and (200) of cubic δ-NbN, respectively.

3.4.2 Compositional uniformity and crystal structure

To analyze the chemical composition of the deposited films, we used a Hiden Analytical MAXIM SIMS equipped with a cesium source and with a spot diameter of 20 µm. The 20 nm thick NbN films were sputtered with a rate of 5 nm/s and the composition of the sputtered material was analyzed using a time-of-flight analytical detector. We traced the most relevant elements present in the films: Nb, N, C, and O. Plateaus in the recorded depth profiles indicate the deposited thin film and were used to extract the intensity of each element. The resulting counts of carbon, nitrogen, and oxygen were normalized with respect to the niobium count at each position to compensate for slight variations of the primary ion current during the measurement. Niobium was chosen as a reference because of two reasons. First, we assume a constant niobium content over the whole area because of the self-limiting ALD process and the saturation of the surface with TBTDEN. In previous studies, we found that the main impact on film composition stems from the plasma step [171, 172]. Second, the sputter rate of niobium should be less dependent on the surrounding chemical bonds because of its higher atomic weight as opposed to carbon, nitrogen, and oxygen. Because the ionization probability of each element is different, the curves of N and C were normalized and absolute values cannot be compared between the elements.

In Fig. 3.7a, the normalized intensities of nitrogen and carbon are depicted. It can be seen that the nitrogen concentration slightly varies over an area of 4-in. diameter, whereas there are larger deviations further from the center. The concentration of nitrogen compared to niobium decreases in the range from ±50 mm to ±75 mm in all directions. A similar behavior is observed for carbon. These declining concentrations indicate a niobium-rich composition at the edges of a 6-in. area.

The presence of oxygen was recorded as well but proved to be widely fluctuating and not suited to evaluate the oxygen content within the films. The penetration depth of the impinging Cs^+ primary ions is in the range of the NbN film thickness of 20 nm. When the surface is sputtered, there is therefore an influence of the SiO_2 layer below (see layer stack in Fig. 3.5e), which is assumed to have the biggest effect on the measured oxygen counts. A reliable measurement of the oxygen concentration would be possible by using pure silicon substrates and removing the native oxide surface layer on the substrate immediately before the deposition. Nevertheless, previous analyses have disclosed a considerable portion of oxygen impurities in the ALD-NbN films [171, 172].

The changes in crystallinity and lattice parameter across the deposition area were evaluated using a PANalytical X'Pert Pro x-ray diffractometer equipped with a Cu K_α source. Because of the small film thickness, the samples were mainly analyzed in grazing incidence ($\omega = 0.4°$) with a parallel beam setup instead of standard Bragg-Brentano configuration. The scan time in the 2Θ-range (33 to 44)° was kept constant at 40 min for each sample.

In Fig. 3.7b, the diffraction peaks of three samples along the x-axis are given. In the center ($x = 0$ mm), two peaks at 35.49° and 41.34° are detected and can be assigned to the orientations (111) and (200) of δ-NbN, respectively. The full width at half maximum (FWHM) of both peaks is rather large (1.02° and 0.84°), indicating a nanocrystalline film structure and a distinct lattice parameter spread [172]. Using the Scherrer equation, crystallite sizes of (9 to 11) nm can be estimated.

At position $x = 40$ mm, the peaks shift to slightly larger angles, but the FWHMs (0.98° and 0.88°) remain nearly constant. Further from the center at $x = 60$ mm, there is a more distinct shift to larger angles, with the peaks at 36.08° (111) and 41.91° (200). Notably, the peak intensity of the (111)-orientation is only slightly dependent on the sample position, while the intensity of (200)-peaks is strongly reduced toward the edge. The disproportionally strong decrease of [100]-crystallites also becomes apparent from the peak-area ratios given in Tab. 3.2.

Within the 2Θ-range (20 to 80)°, five diffraction peaks from cubic NbN were recorded. Only in the center ($x = 0$ mm), one very small peak at ≈51° from another phase was observed, either from Nb_4N_5 or from niobium oxynitride. Nevertheless, previous investigations revealed the presence of impurities within the deposited films [171, 172]. Therefore, the coexistence of amorphous phases of niobium oxides, niobium carbides, and niobium oxynitrides might be possible. The amorphous content, however, should be very small because diffractograms recorded in Bragg-Brentano configuration do not show the occurrence of characteristic "amorphous humps".

Table 3.2: Peak position, FWHM, and peak-area ratio of recorded XRD peaks. For the area ratios of the (111)- to (200)-peaks, we subtracted the baseline of each curve and integrated the peaks in the ranges (34 to 37.5)° and (40 to 43)°, respectively.

	peak position (°)		FWHM (°)		peak-area ratio
x (mm)	(111)	(200)	(111)	(200)	(111)/(200)
0	35.49	41.34	1.02	0.84	0.81
40	35.61	41.45	0.98	0.88	1.03
60	36.08	41.91	0.96	1.05	1.57

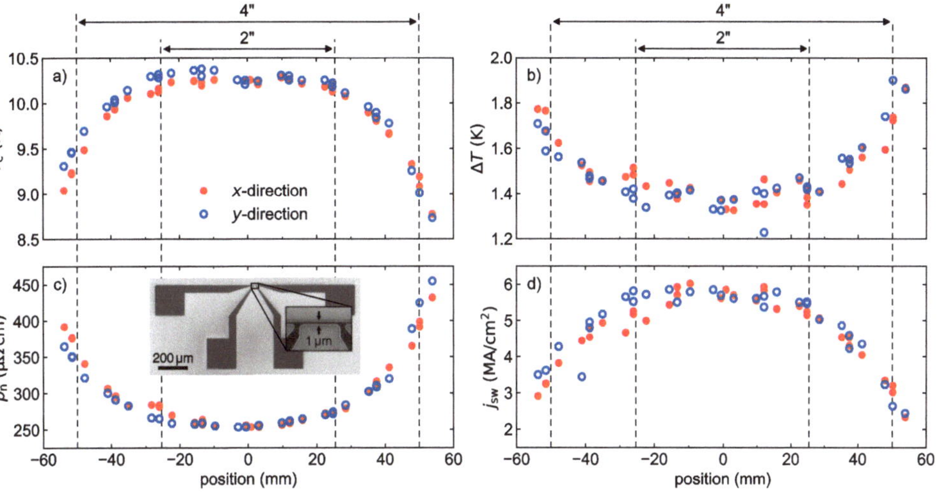

Figure 3.8: Distribution of critical temperature (a), superconducting transition width (b), residual resistivity at 25 K (c), and switching current density at 4.2 K (d) over the sample's position on the wafer along the x- and y-axis. The measurements were conducted on microbridges of 1 μm width and 6.1 nm thickness (see the scanning electron micrograph in the inset). The vertical dashed lines illustrate wafer edges of the denoted sizes.

3.4.3 Spread of superconducting properties

The superconductivity parameters were measured on a total of 68 microbridges (four per chip) distributed along the x- and y-axis of the examined area. For T_c, ΔT, and ρ_n, the temperature-dependent resistance was recorded, while current-voltage curves at 4.2 K were used to get switching current density j_{sw}. In addition, the temperature dependence of j_{sw} between T_c and 4.2 K was measured on several samples. For ρ_n and j_{sw}, we took cross-sections obtained by SEM and AFM measurements.

The spread of these superconductivity metrics is displayed in Fig. 3.8. Across a 2-in. diameter, only minimal variation is visible and the electrical parameters approximately amount to $T_c = 10.3\,K$, $\Delta T = 1.38\,K$, $\rho_n = 260\,\mu\Omega\,cm$, and $j_{sw} = 5.7\,MA/cm^2$. Further distant from the center, the film quality is degrading by varying degree of each property. At $x, y = \pm 50\,mm$ (corresponding to the edges of a 4-in. wafer), the resistivity is on average increased by 53 % compared to the center. Similarly, the switching current density decreases by 45 %. The critical temperature is shown to be more stable over the deposition area compared to ρ_n and j_{sw} and decreases by 10 % only. This is expected since T_c is mainly determined by any contiguous, high-quality region along the microbridge. To evaluate film homogeneity, the transition width is better suited because it encompasses the spread of T_c of various regions in the film. From the center to the 4-in. wafer edge, ΔT increases by ≈26 %. This points to a degrading homogeneity of different grains and regions of the thin film.

Within the uniform area in the center, the critical temperature is in line with the measurements in Section 3.2 for a thickness of 6 nm but on sapphire and quartz substrates. From the other metrics, however, ALD-NbN on thermal oxide seems to be a slightly better superconducting material, with a 20 % higher j_{sw} and ≈20 % lower values of ρ_n and ΔT, respectively. Compared to properties of unpatterned, slightly thicker ALD-NbN on SiN_x reported by Cheng *et al.* [225], we obtained very similar values for ρ_n, while T_c is (2 to 3) K higher and ΔT about 0.3 K lower in our case. Reasons for this discrepancy could be different deposition temperatures (380 °C vs. 300 °C), plasma gas compositions (pure H_2 vs. mixed N_2/H_2), and substrate materials (thermally oxidized Si vs. LPCVD-grown SiN_x).

We analyzed whether the decreasing j_{sw} toward the wafer edges can be entirely explained by the change of T_c and ρ_n over the same distance. This can be accomplished by comparing the ratios of the measured switching currents density j_{sw} to the theoretical depairing current density $j_{dep}(0) = 0.74\Delta(0)^{3/2}/(e\rho_n\sqrt{\hbar D})$, with Eq. 31 in Ref. [226], the energy gap $\Delta(0) = 2.05k_BT_c$, and the density of states $N(0) = 1/(2e^2\rho_n D)$.

For the micrometer-wide bridges like in our case, the barrier for vortex entry is considerably reduced compared to (20 to 100) nm wide bridges (see also results and discussion in Section 3.2.2). This decreases the measurable j_{sw} for $T/T_c \lesssim 0.8$. Although this effect should be similar for all bridges with the same dimensions, it introduces an uncertainty because any edge defect can reduce the entry barrier further in a hardly predictable way. This is why we did not take the measured j_{sw} values at 4.2 K directly but instead recorded $j_{sw}(T)$ curves of each sample and fit the experimental data with

$$j_{fit}(T) = j_{fit}(0) \cdot \left[1 - (T/T_c)^2\right]^{3/2} \tag{3.8}$$

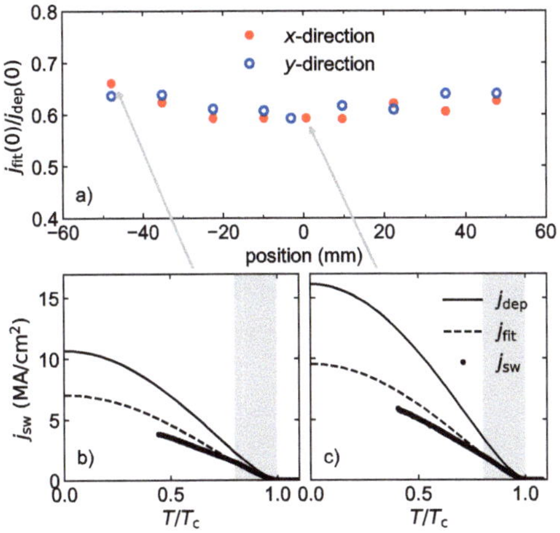

Figure 3.9: (a) Distribution of the ratio of fit current density j_{fit} to theoretical depairing current density j_{dep} at 0 K over the wafer area. (b), (c) Exemplary $j_{\text{sw}}(T)$ curves at $x = -47.8$ mm (b) and $x = 0.7$ mm (c). The dots represent j_{sw} measured on bridges of 1 µm width, the dashed curves denote fits by Eq. 3.8 to the measured data in the range of $0.8 < T/T_{\text{c}} < 1.0$ (indicated by the shaded area), and the solid line shows the depairing current density j_{dep}, obtained from $j_{\text{dep}}(0)$ and temperature dependence in Eq. 3.8.

in the temperature range $T \geq 0.8T_{\text{c}}$ (Bardeen approach, see Eq. 5.26 in Ref. [227]; compare to Eq. 3.5 in Section 3.2.2). In this range close to T_{c}, the coherence length diverges and j_{sw} is mostly determined by the depairing mechanism [228, 229]. The fit parameter $j_{\text{fit}}(0)$ is then compared to the theoretical value of $j_{\text{dep}}(0)$, calculated using the respective T_{c} and ρ_{n} at each x-/y-position. The coefficient $D \approx 0.3\,\text{cm}^2/\text{s}$ is taken from previous measurements on NbN of the same thickness, albeit on a crystalline quartz substrate (see Fig. 3.2e), and is assumed to be constant over the wafer area.

As shown in Fig. 3.9a, the ratio $j_{\text{fit}}(0)/j_{\text{dep}}(0)$ amounts to $\approx$0.6 in the center. This value is close to the ratio shown previously in Fig. 3.3 for $w = 100$ nm. Over the examined area, the current ratio is almost constant with a maximum deviation of 10 % compared to the center. A constant ratio $j_{\text{fit}}(0)/j_{\text{dep}}(0)$ means that $j_{\text{fit}} \propto T_{\text{c}}^{3/2}/\rho_{\text{n}}$ and the measured drop of j_{sw} is in line with the changes in T_{c} and ρ_{n} at each position. The fact that the ratio slightly increases toward the edges is most likely because the assumption of constant diffusion coefficient D is not applicable. Instead, D should decrease toward the edges as it is expected for higher resistivity [156], and therefore, the ratio $j_{\text{fit}}(0)/j_{\text{dep}}(0)$ should be smaller than calculated.

The degradation in film quality starting at $\approx$25 mm from the center can not be explained by morphological changes because both the film thickness and surface roughness are fairly uniform within a 4-in. area ($|x, y| \leq 50$ mm). A decreasing nitrogen content in the film relative to niobium was observed, but this could only explain a change of the superconducting properties further distant from the center (>50 mm).

Instead, XRD measurements suggest changes in the film structure as a possible reason for this drop. The diminishing peak intensity, especially of the (200)-oriented grains, signifies a declining portion of cubic NbN toward the wafer edges. The detected 2Θ-peak shift with increasing distance from the center corresponds to a decrease in the lattice parameter from 4.37 Å to 4.31 Å. According to the influence of this lattice parameter reduction on the superconducting properties described by Linzen *et al.* [172], a decrease in T_c by (2 to 3) K is expected.

The maximum temperature deviation over the substrate table is specified to be ±2 °C. Hence, we do not expect inhomogeneities in deposition temperature to be a factor contributing to film degradation toward the edges.

In previous studies, we found plasma parameters to be crucial for final film composition and electrical properties. A decrease in hydrogen pressure resulted in higher T_c and j_{sw} values [171]. In addition, low hydrogen pressure led to an increase in plasma radicals near the substrate surface [169]. This suggests that in the downstream region, the remote plasma could rather act as an active plasma. Moreover, the low pressure additionally prevented interactions of radicals within the downstream due to the increased mean free path, leading to a rather directional plasma exposure. The plasma source in our setup has a diameter of 75 mm and is concentrically aligned above the substrate holder (see sketch in Fig. 3.5a). Therefore, the directional plasma is expected to have its highest energy impact on the substrate within a region of approximately ±40 mm. This is consistent with our results indicating that the film composition, the lattice parameter, and the superconducting properties are significantly altered at distances ≥40 mm from the center in all directions.

Based on these results, the large-area homogeneity of ALD-NbN might be improved by using a larger-diameter plasma source and increasing the ICP-to-substrate table distance (see also results from Section 3.3). This would result in an extended area of uniform plasma distribution and reaction with the deposited precursor.

3.5 Summary

In this chapter, the deposition of polycrystalline, superconducting NbN thin films using ALD was investigated and evaluated in view of its suitability for the fabrication of SNSPDs.

Table 3.3: Summarized parameter spread of ALD-NbN.

		change (%) over diameter	
parameter	center value	2 in.	4 in.
d	6.1 nm	<5	<5
R_q	0.82 nm	<3	<6
T_c	10.3 K	−1	−10
ΔT	1.38 K	+5	+30
ρ_n	260 µΩ cm	+6	+50
j_{sw}	5.7 MA/cm^2	−8	−45

Basic transport measurements on films of (3 to 21) nm thickness on sapphire and quartz substrates revealed mostly constant properties down to a thickness of ≈5 nm, with a residual resistivity ρ_n below 400 µΩ cm and a critical temperature T_c above 10 K for films on sapphire. At $d \leq 5$ nm, the film quality is rapidly degrading, likely owing to the shrinking superconducting core between the film surface and the substrate interface. Qualitatively, films on quartz exhibit a similar thickness dependence, but overall slightly "worse" superconducting properties compared to sapphire (although the difference diminishes for thicker films). With an electron density of states $N(0) \propto (\rho_n D)^{-1}$ comparable to sputtered films, the heat capacity ratio C_e/C_{ph} of ALD-NbN should also be similar. Compared to reference values, the diffusion coefficient of ALD-NbN is rather small at $D = 0.32\,\mathrm{cm^2/s}$ for 10 nm on sapphire, which is advantageous for SNSPDs.

Further material properties relevant for SNSPDs were obtained by measurements on micro- and nanobridges of width (0.1 to 10.0) µm to extract the temperature- and width-dependent switching current density j_{sw}. Although the j_{sw} values of ALD-NbN bridges were smaller compared to sputtered ones in absolute terms, in conjunction with the higher ρ_n and lower T_c of ALD-NbN, the most crucial switching to depairing current ratio (with I_{dep} obtained from the model by Kupriyanov and Lukichev [218]) is very similar at values close to $I_{sw}/I_{dep} = 0.6$. With a rather large ρ_n and smaller T_c and j_{sw}, the investigated ALD-NbN films are "worse" superconductors compared to NbN film sputtered at high temperatures [85], but meet the specific requirements of SNSPDs very well (high ratio I_{sw}/I_{dep}, small D, moderate T_c).

Continuing previous investigations of the influence of various plasma parameters on the film properties [171], the interaction between the plasma and the substrate surface was modified by varying the distance between the sample and the inductively coupled plasma source. This resulted in gradually worse properties (higher ρ_n, lower T_c and j_{sw}), suggesting that a reduction of the plasma power might actually be beneficial and worth to pursue.

The main goal of employing ALD for the deposition of NbN, enabling highly uniform thin films over large areas, was investigated by measuring the spread of various film parameters over the area of a 6-in. wafer. All observed film properties including the most decisive switching current density are near-constant over a diameter of 50 mm. Further from the center, at a distance of ±50 mm, we observe larger variations of the electrical properties: +50 % (ρ_n), −10 % (T_c), +30 % (ΔT), and −45 % (j_{sw}), while the film thickness and surface roughness are mostly unchanged (see also summary in Tab. 3.3). XRD measurements suggest that this change of parameters is most likely caused by a shrinking crystalline portion of cubic NbN together with a decrease in the lattice constant. Correlations with the lateral dimension of the downstream region of the plasma suggest that the film uniformity could be extended to larger areas by increasing the diameter of the plasma source.

Overall, the presented ALD-NbN thin film properties indicate a high suitability for SNSPDs, allowing for large-area detector arrays on a 2-in. wafer with atomic-level thickness control and potentially increasing the fabrication yield compared to conventional reactive magnetron sputtering.

4 ALD-NbN SNSPDs

The transport measurements of ALD-NbN in the previous chapter revealed promising properties for the use in nanowire detectors. In this chapter, actual ALD-NbN-based SNSPDs are fabricated and characterized under optical illumination.

In Section 4.1, the detector layout and sample preparation with the focus on the patterning process is described. This is followed by the experimental setup in Section 4.2. The results of the optical detector characterization in terms of bias, photon-rate, and spectral dependence of the detector response are presented together with a discussion of the film-thickness influence in Section 4.3.

4.1 Detector design and fabrication

The optical characterizations were carried out on detectors from ALD-NbN films deposited using the process parameters as specified in Section 3.1. To investigate the influence of the film thickness on the detector properties, both 5 nm and 10 nm thick detectors were fabricated. For both depositions, 10 mm × 10 mm *R*-plane sapphire served as substrate. In the following, the layout and patterning process of the detectors is introduced.

4.1.1 Detector design

An overview of a single detector chip together with zooms on the nanowire is shown in Fig. 4.1. Nine of these devices were placed on a 10 mm × 10 mm area for parallel fabrication.

Straight nanowires with a width of ≈100 nm and a length of 5 µm define the active area of the devices. Although this basic nanowire geometry is not suited for efficient photon coupling, it serves the purpose of characterizing the fundamental single-photon response of ALD-NbN-SNSPDs by minimizing patterning defects commonly found in large-area detectors. To reduce the current crowding effect (see Section 2.3.1 and Ref. [82]), the nanowire is connected by small tapers on each end, as seen in the inset of Fig. 4.1b.

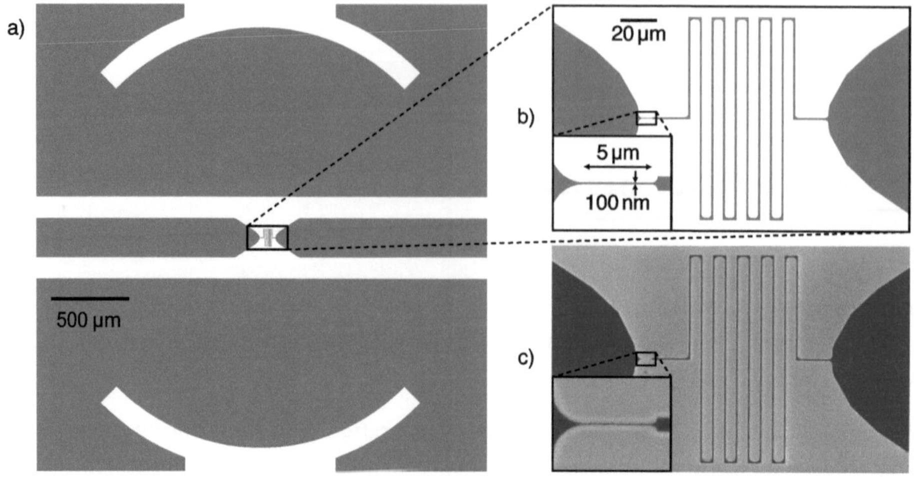

Figure 4.1: Design of the SNSPD chip. (a) Detector chip with a size 3 mm × 3 mm. The SNSPD and inductor in the center are connected to CPWs on both sides. The gray areas denote the NbN layer. (b) Zoom of chip center with the straight nanowire (inset) and a series inductor. (c) SEM micrograph of the fabricated device.

Since the small kinetic inductance of the short nanowires would lead to detector latching (described in Section 2.2.3 and later investigated in Section 5.2.1), a 1 µm wide and 1 mm long inductor was placed in series with the nanowire. With 1000 squares, this adds an inductance of about 46 nH for $d = 10$ nm. The structure much wider than the nanowire ensures that, at currents below the switching current of the nanowire, the inductor itself is not sensitive to photons but only the nanowire (at least in terms of the SNSPD detection mechanism).

The nanowire and inductor are connected to coplanar waveguides (CPW) from the same NbN film for bias and readout. The 250 µm wide, inner conductor was designed wide enough to enable conveniently connecting the CPW by aluminum wire bonds to a printed circuit board (PCB) for readout. To set an impedance of 50 Ω, the inner conductor of the CPW is separated by 140 µm wide gaps to the ground planes on each side. Both ground planes feature integrated, etched rings with an outer diameter of 3 mm. These markers are used to facilitate the manual positioning of the chip in the detector housing. At a minimum distance of 700 µm from the CPW gaps (five times the gap size), these structures have a negligible impact on the CPW impedance.

4.1.2 Detector patterning

The detectors were patterned out of NbN films via electron-beam lithography (EBL) and Ar-ion milling.

The EBL comprised three separate steps. First, the nanowires and alignment marks (not visible in Fig. 4.1) were defined in a negative-tone process. The chip was spin-coated with poly (methyl methacrylate) (PMMA, 950k) and baked at 150 °C for 5 min to get a resist thickness of ≈80 nm. The structures were then written by EBL at 15 kV acceleration voltage, a beam current of 14 pA, and a high dose of 10 mC/cm^2, which causes cross-linking of the exposed polymers and PMMA effectively acting as a negative-tone resist [230]. Nanowires fabricated by this process were found to have smoother edges and enhanced superconducting properties due to a more homogeneous cross-section [87]. After exposure, the resist was developed by submersion in acetone (1 min) and rinsing in isopropanol (IPA; 30 s). In the second and third lithography step, the coarse and mid-size structures were patterned using a standard positive-tone process to avoid the impractically long exposure times which would be necessary for the larger areas exposed with the same high dose as the nanowires. These subsequent writing steps were aligned to the nanowires using the previously written alignment marks. To write the coarse structures on the scale of ≈100 µm including the CPWs, chip-positioning rings, and chip labels, the same type of PMMA resist was spin-coated with the same parameters. For illumination, however, a larger beam current of 4.5 nA and a much lower dose of 130 $\mu C/cm^2$ were used, corresponding to the regular positive-tone process for PMMA. The exposed resist was developed in methyl isobutyl ketone (MIBK):IPA 1:3 for 35 s before rinsing in IPA for 10 s. Lastly, without applying a new resist layer, the mid-size patterns on the scale of ≈1 µm, namely the inductors, were defined in a third exposure at the same dose as before but at a lower beam current of 120 pA. The lower current allowed for a higher resolution and a lower edge roughness of the micrometer-wide inductors than what was necessary for the coarse structures. After lithography, the resist was again developed in MIBK:IPA 1:3 but for a slightly shorter time of 25 s. The final PMMA structure comprised the nanowire, written in the negative process, as well as the CPW and inductor, both written using a positive process.

Following the lithography and development, the patterned resist structures were transferred to the NbN layer by Ar-ion milling. For this process, the samples were placed in a vacuum chamber below an RF-coupled ion source on a rotating sample holder to ensure uniform etching. The argon pressure was set to $\approx 2.8 \times 10^{-4}$ mbar before igniting the plasma. The generated ions were accelerated toward the sample by the electrostatic field of two electrode grids which were biased by DC voltages of +200 V (inner electrode) and −250 V (outer electrode). The RF plasma power was adjusted to achieve a constant current at the positive electrode grid of 90 mA, which is proportional to the ion-beam current. Using these parameters, the etching rates of PMMA and NbN were previously determined to be around 7 nm/min and 3 nm/min, respectively. The cross-linked PMMA (negative-tone EBL) is more resistant to the ion impact and was found to have an etch rate similar to NbN. For the 5 nm and 10 nm thick samples, etch times of 6 min and 10 min were chosen, respectively. The etch times were much longer than required according to the etch rates in order to ensure that the NbN layers were etched through. Because of the single-layer detectors, an overetching into the substrate had no detrimental effect on the samples. Moreover, there is only minimal etching into the hard sapphire substrate. To prevent overheating of the NbN layer due to the ion-beam impact, the sample holder was cooled to ≈0 °C. As an additional procedure to reduce the heat load, the sample was not etched continuously but in intervals of 30 s with 30 s long periods in between during which the shutter between the ion source and the sample was closed.

After etching, the remaining resist was removed by submersion in acetone, and the 10 mm × 10 mm substrates were each diced into nine individual chips. Because the cross-linked PMMA is insoluble in acetone, it was left on the nanowires as a protective layer. An SEM micrograph of the fabricated SNSPD is shown in Fig. 4.1.

4.2 Experimental setup

For the optical characterization, the detectors were mounted in a LHe-cooled dipstick cryostat equipped with an optical fiber and both RF and DC lines. All measurements were conducted at 4.2 K. The detectors were biased by low-noise voltage source with low-pass RC filters (≈72 Hz edge frequency), connected to the DC port of a bias tee. The detector signal at the RF output of the bias tee was amplified at room temperature by a two-stage amplifier with a total gain of ≈67 dB. Depending on the applied bias current, the resulting pulse height after amplification was in the range (200 to 900) mV. A 32 GHz real-time oscilloscope and a pulse counter for count rates up to 300 MHz were used for readout.

For illumination, the output of various light sources was coupled to the detectors at the bottom of the dipstick using a step-index multimode fiber specified for the wavelength range (400 to 2100) nm. The detectors were illuminated from the top, normal to the substrate plane. In order to ensure a uniform illumination, the detectors were placed at a distance of (4 to 5) mm from the fiber tip, resulting in a widely illuminated area with a radius of about 1 mm. For measurements in dependence of the bias current and optical power, we used an LED at 385 nm and a laser diode at 658 nm wavelength. The output power of both sources could be swept over a broad range by the applied current. For spectral measurements in the wavelength range (400 to 2000) nm, a broadband halogen lamp in conjunction with a Czerny-Turner monochromator were used. The monochromator has two diffraction gratings with different blaze wavelengths: one with 1200 G/mm (grooves per millimeter) for the wavelength range (400 to 1100) nm, and another one with 600 G/mm for (1100 to 2000) nm. In order to block light of higher-order diffraction (e.g., at 400 nm if the grating is set to 800 nm), two longpass filters were used: one with a cut-on wavelength of 750 nm for the wavelength range (800 to 1400) nm, and one with a cut-on wavelength of 1050 nm for (1400 to 2000) nm.

All of the light sources were calibrated using a powermeter equipped with Si- and InGaAs-based photodiodes for the wavelength range (350 to 1100) nm and (800 to 1700) nm (Thorlabs S150C and S154C), respectively. The calibration was conducted in the detector plane at the bottom of the dipstick in order to take into account any wavelength-dependent loss contribution from the vacuum fiber feedthrough, the fiber itself, and the output of the stripped fiber.

4.3 Optical detector characterization

This section contains parts from previously published content:
[209] E. Knehr, A. Kuzmin, D. Y. Vodolazov, et al., "Nanowire single-photon detectors made of atomic layer-deposited niobium nitride", Superconductor Science and Technology ***32****, 125007 (2019).*
Permission to include adapted text was obtained from the authors. The films were deposited by M. Ziegler and S. Goerke at Leibniz-IPHT. E. Knehr designed, patterned, and characterized the detectors and analyzed the results with contributions from A. Kuzmin, K. Ilin, and D. Vodolazov.

Before photon illumination, the basic superconducting properties of the fabricated detectors were measured. The results are summarized in Tab. 4.1. While the values for $d = 5$ nm are in line with the results from Fig. 3.2, the 10 nm sample has slightly smaller values for both ρ_n and T_c. Ratio I_{sw}/I_{dep} was obtained via Eq. 3.5 (with diffusion coefficient D from Fig. 3.2e and $\beta_0 = 2.05$) and is consistent with the result of Fig. 3.3b for the same thickness of 10 nm, whereas a smaller ratio $I_{sw}/I_{dep} = 0.4$ is found for the thinner sample.

4.3.1 Bias and photon-rate dependence

The bias-dependent count rates for different film thicknesses, photon wavelengths, and optical power levels are shown in Fig. 4.2. As summarized in Section 2.3.1, SNSPDs exhibit three

Table 4.1: Superconducting properties of SNSPDs with 5 nm and 10 nm thickness. Values of I_{sw}, j_{sw}, and I_{sw}/I_{dep} are obtained at 4.2 K.

d (nm)	ρ_n (μΩ cm)	T_c (K)	I_{sw} (μA)	j_{sw} (MA/cm^2)	I_{sw}/I_{dep}
5	385	10.1	15.1	3.02	0.40
10	309	9.0	43.0	4.3	0.51

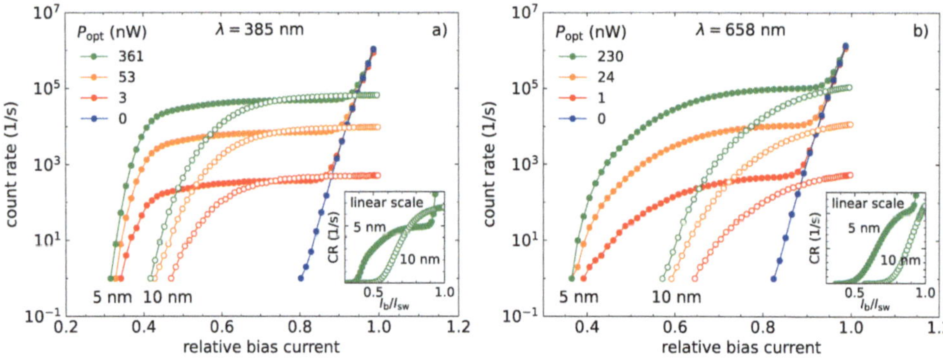

Figure 4.2: Count rate over relative bias current I_b/I_{sw} for illumination with photons at 385 nm (a) and 658 nm (b) wavelength and different optical power levels. The curves at 0 nW correspond to dark-count rates. (Insets) Respective plots on the linear scale for the highest power level.

distinct regimes dependent on the applied bias current, which is reflected in the experimental results. At low currents, e.g., below $0.4I_{sw}$ for $\lambda = 385$ nm and $d = 5$ nm, there is an exponential increase in count rate with the bias. The strong bias dependence is associated with a probabilistic or fluctuation-assisted photon detection. At higher currents, the curves saturate (in logarithmic scale) which is attributed to the deterministic detection regime.

At $\lambda = 385$ nm (Fig. 4.2a), all measured curves exhibit saturating count rates. As indicated by the prolonged plateau in CR(I_b), the thinner sample shows a deterministic regime over a wider bias range. The saturating curves in linear scale (see inset) signify near-unity internal detection efficiencies at I_b close to I_{sw} at this photon energy [46]. For the longer wavelength ($\lambda = 658$ nm, Fig. 4.2b), the plateaus are less pronounced (5 nm) or not existing anymore (10 nm). This is again better illustrated by the curves on the linear scale (inset). Comparing curves of both thicknesses at the same optical power, the level of saturation (extrapolated in the case of $d = 10$ nm and $\lambda = 658$ nm) is largely independent of the detector thickness, further emphasizing the near-unity internal efficiency. Small differences in the saturation level might come from the fact that both detectors were measured in separate cool-downs

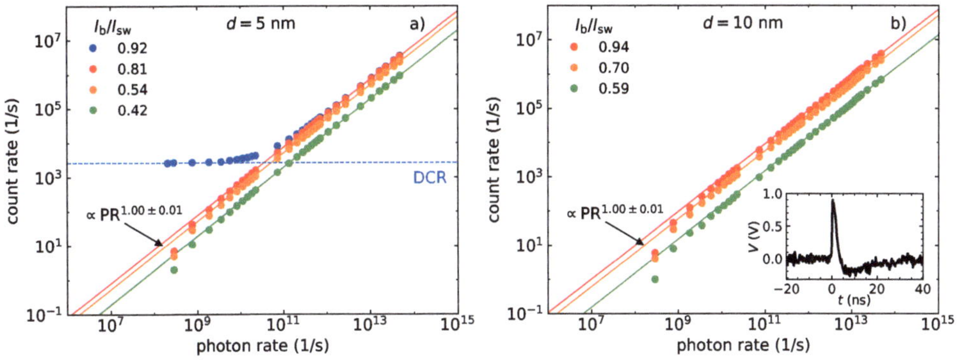

Figure 4.3: Count rate over incoming photon rate (λ = 385 nm) for different bias currents of 5 nm (a) and 10 nm (b) thick detectors. The solid lines are fits according to CR $\propto$ PR^n to the respective data. The dashed horizontal line in (a) indicates the DCR at $I_b = 0.92 I_{sw}$. (Inset) Basic output pulse of the 10 nm thick detector.

with potential deviations in the optical coupling, i.e., different photon rates impinging the detector at the same optical input power.

The dark-count rate (DCR) was measured by blocking the fiber feedthrough of the dipstick. Dark counts were found to appear at currents above $0.8 I_{sw}$ and to grow exponentially with increasing current. This, however, was only observed for d = 5 nm, whereas the 10 nm sample did not exhibit an increasing DCR above a current-independent noise level of ≤3 Hz up to I_{sw}.

As pointed out in Section 2.3.1, the operation temperature has a profound influence on the photon sensitivity of SNSPDs. Although the measurements were conducted at $T = 4.2\,\text{K} < 0.5 T_c$ and the short nanowire is beneficial for detecting low-energy photons (minimal current crowding, small probability of nonuniformities), it is obvious that the operation conditions at hand do not allow for an efficient photon detection with saturated CR(I_b) curves at longer wavelengths, i.e., in the infrared range. The temperature influence on the shape of CR(I_b) is, e.g., shown in the extended data Fig. 1a of Ref. [8], which demonstrates much longer count-rate plateaus when lowering the temperature from 4 K to even 2 K.

Figure 4.3 depicts measured CR(PR) curves at λ = 385 nm. It is seen that the count rates increase with the incoming photon rate PR as CR $\propto$ PR^n with an exponent $n = 1.00 \pm 0.01$. This applies to both thicknesses and all I_b/I_{sw} ratios corresponding to different bias regimes. Similar results were also obtained for a photon wavelength of 658 nm. The linear dependence CR $\propto$ PR over many orders of magnitude is a strong indication of single-photon detection [4, 231]. A saturation in count rate approaching the maximum count rate of the detectors could not be observed because of the weak optical coupling from the fiber to the active area and,

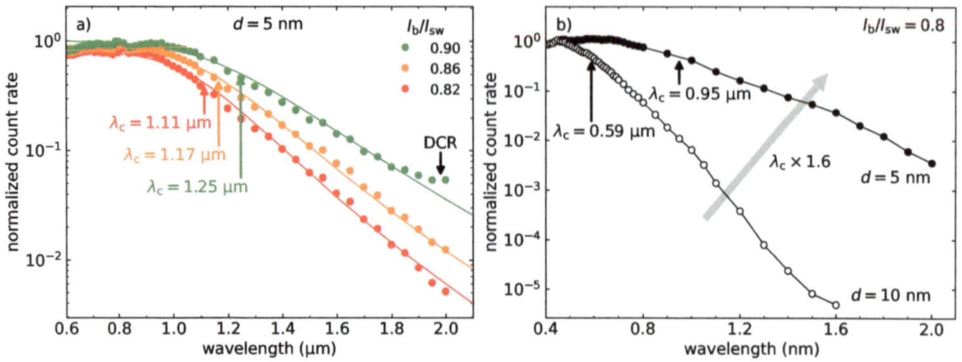

Figure 4.4: Spectral dependence of the normalized count rate. (a) Normalized count rate over photon wavelength at different bias currents for the 5 nm thick detector. The solid lines are fits via Eq. 2.12. Denoted cut-off wavelengths are obtained from the fits. (b) Comparison of wavelength-dependent count rate of the 5 nm and 10 nm thick detectors at the same current ratio $I_b/I_{SW} = 0.8$.

consequently, heating of the sample which suppressed I_{sw} and therefore limited I_b. As seen in Fig. 4.3a for $d = 5$ nm, the count rate at $I_b = 0.92 I_{sw}$ saturates to DCR at low photon rates. In all cases, a divergence between the experimental data and the linear progression is visible at low photon rates, which is likely due to the calibration precision of the optical powermeter. The specified minimum power of about 100 pW corresponds to PR $\approx 2 \times 10^8$ Hz (3×10^8 Hz) at $\lambda = 385$ nm (658 nm).

4.3.2 Spectral dependence

Spectral measurements were conducted by sweeping the monochromator's output wavelength in steps of 10 nm to 100 nm and recording count rates at each wavelength. The output power of the monochromator was strongly dependent on the wavelength. In order to keep the detector in the same detection regime during spectral measurements, the count rate was kept at values below 1×10^4 Hz by manually adjusting the input slit width of the monochromator accordingly. Furthermore, because of different gratings and filters employed over the whole wavelength range, several wavelength sweeps with partially overlapping data points were conducted. These were then used to stitch the count rates and the optical power levels starting at the shortest wavelength. All curves were normalized to the count rate at short wavelengths.

In Fig. 4.4a, the normalized count rate is plotted as a function of the incident photon wavelength from 0.6 µm to 2.0 µm. Below ≈1.0 µm, there is a plateau in the recorded count rates.

This indicates a deterministic detector response and, thus, matches the results of Fig. 4.2. The fact that the saturation level slightly increases with higher count rates likely implies that an increased portion of the nanowire is active, which fits to the near-saturated CR(I_b) curves in linear scale (and therefore $\eta_{int} \lesssim 1$) in the insets of Fig. 4.2.

Above $\lambda \approx 1\,\mu m$, the count rates at a fixed ratio I_b/I_{sw} drop rapidly with the photon wavelength. To obtain the cut-off wavelength λ_c, each curve was fit by Eq. 2.12 with λ_c, slope p, and a pre-factor denoting the saturation level as fitting factors. At the highest bias current ($I_b = 0.90 I_{sw}$), we obtained $\lambda_c = 1.25\,\mu m$ (see the annotations in Fig. 4.4a). Similar measurements on the 10 nm sample resulted in $\lambda_c = 0.88\,\mu m$ at $I_b = 0.98 I_{sw}$. It should be noted that there are different ways to define the cut-off wavelength. Besides the fit used here, an approximation of the dependence by two straight lines and defining λ_c at the intersection is sometimes used as well [85]. In general, the latter definition leads to slightly smaller values of λ_c. As it can be seen, increasing the bias current goes along with a longer cut-off wavelength and a shallower drop in count rate above λ_c. Dependence $\lambda_c(I_b)$ is described by various detection models [40, 46, 48, 232] and was also probed experimentally [79, 80, 85]. By increasing the bias current, the energy barrier for vortex entry is reduced and the superconducting state becomes unstable at lower photon energies [42, 95, 233]. Also, larger portions of the nanowire contribute to photon detection at higher bias currents (related to the position-dependent detection current pointed out in Section 2.3.1).

Similarly, CR(λ) depends on the film thickness. As illustrated in Fig. 4.4b, we measured an increase in cut-off wavelength by a factor of 1.6 for the same ratio $I_b/I_{sw} = 0.8$ when reducing the thickness from 10 nm to 5 nm. This is approximately in line with the proportionality $\lambda_c \propto d^{-1}$ since less demand on ratio I_b/I_{sw} is made for detectors with smaller cross-sections (see again Section 2.3.1 and discussion below). The change in λ_c by a factor smaller than the ratio of two in film thickness is at least partly due to the fact that the material parameters also change with the thickness. The smaller diffusion coefficient D (see Fig. 3.2e), smaller energy gap $\Delta \propto T_c$, and larger current ratio I_{sw}/I_{dep} (see Tab. 4.1) of the 10 nm sample all favor photon sensitivity and cause a longer cut-off wavelength. Hence, the difference in λ_c of both detectors is smaller than expected from the ratio in film thickness.

4.3.3 Thickness-dependent dark and photon counts

According to existing SNSPD detection models, both photon counts in the probabilistic regime and dark counts can be explained by fluctuations, e.g., vortex hopping [41, 47], vortex-antivortex-pair (VAP) unbinding [39, 45, 48, 234–236], Fano fluctuations (in the case of photon counts) [72], and local thermal fluctuations [73]. Here, the influence of the film thickness on dark and photon counts is estimated. Because the energy for VAP creation is almost double the energy for a single vortex [41, 233], only single-vortex entry is considered.

According to Eq. 2 in Ref. [233], the energy barrier for vortex entry of superconducting stripes without defects at currents close to the depairing current can be expressed by

$$\mathrm{d}F = \mathrm{d}F_0 \left(1 - \frac{I}{I_{\mathrm{dep}}}\right), \tag{4.1}$$

with current-independent factor

$$\mathrm{d}F_0 = \frac{\Phi_0^2}{2\pi\mu_0\Lambda}. \tag{4.2}$$

Via Pearl length $\Lambda = 2\lambda_1^2/d$, $\mathrm{d}F$ and $\mathrm{d}F_0$ scale linearly with thickness d. The vortex-hopping rate (or DCR) depends on the energy barrier $\mathrm{d}F$ and is estimated by

$$\Gamma \propto \frac{l}{\xi_{\mathrm{GL}}} \nu \cdot \exp\left(-\frac{\mathrm{d}F}{k_{\mathrm{B}}T}\right), \tag{4.3}$$

with the attempt frequency

$$\nu \propto \frac{\Delta}{\hbar}\sqrt{\frac{\mathrm{d}F}{k_{\mathrm{B}}T}}. \tag{4.4}$$

This is in analogy to the theory of phase slips in one-dimensional superconducting wires developed in Refs. [237, 238]. Although this theory assumes a defectless stripe and describes thermally activated fluctuations only close to T_c (within the Ginzburg-Landau model), it was found in Refs. [233, 239] that it is valid also for lower temperatures well below T_c and holds also in the presence of defects with minor modifications. Furthermore, it was expanded to two-dimensional stripes in Ref. [240].

In Eq. 4.3, factor l/ξ_{GL} is the number of independent points for vortex entry along the nanowire and amounts to $\approx$1000 (see the measured ξ_{GL} in Fig. 3.2f). However, because vortex entry is considered to originate predominantly from only a few weak spots of size $\approx\xi_{\mathrm{GL}}$ along the stripe where the energy barrier is lowered, it can be assumed to be on the order of one. Moreover, we take a value of 1×10^{12} Hz for $\Delta/\hbar$ in Eq. 4.4 as a typical order of magnitude for NbN with a critical temperature of 10 K.

Using ρ_n and T_c from Tab. 4.1 together with λ_l from Eq. 3.1, the Pearl lengths of the 5 nm and 10 nm detectors amount to $\Lambda \approx 144$ µm and 65 µm, respectively. With Eq. 4.2, this corresponds to $\mathrm{d}F_0/k_{\mathrm{B}}T \approx 65$ for 5 nm and $\mathrm{d}F_0/k_{\mathrm{B}}T \approx 144$ for 10 nm. From Eq. 4.1, the different $\mathrm{d}F_0$ should go along with a difference in switching current I_{sw} relative to I_{dep}. For example, if we assume

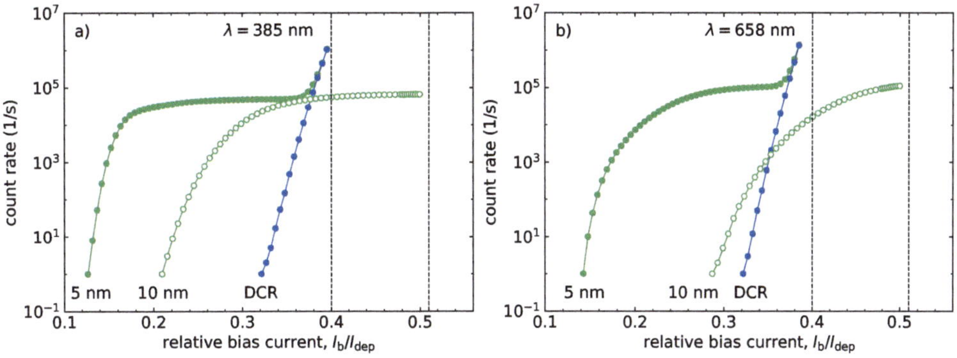

Figure 4.5: Count rate over relative bias current I_{sw}/I_{dep} for illumination with photons at 385 nm (a) and 658 nm (b) wavelength (equivalent to Fig. 4.2 but with the current scaled to units of I_{dep} instead of I_{sw}). The vertical dashed lines denote the respective switching current.

the vortex-hopping rate at which the superconducting stripe switches to the resistive state to be $\Gamma \approx 10\,000$ Hz ($\mathrm{d}F/k_BT \approx 20$), we obtain switching currents of $I_{sw} \approx 0.69 I_{dep}$ ($d = 5$ nm) and $I_{sw} \approx 0.86 I_{dep}$ ($d = 10$ nm).

In the presence of edge defects, the current-dependent energy barrier may differ from the simple dependence in Eq. 4.1, as shown in Refs. [233, 239]. Nevertheless, assuming otherwise constant material properties, dF is directly proportional to d and we expect higher relative switching currents I_{sw}/I_{dep} (not just absolute values of I_{sw}) for the thicker detector. As it is seen from the values in Tab. 4.1, we indeed observe an increase in I_{sw}/I_{dep} by 20 % when going from 5 nm to 10 nm. This is also visualized in Fig. 4.5 which shows the data from Fig. 4.2 but with the bias current scaled to units of I_{dep}. Besides the larger ratio I_{sw}/I_{dep}, the thicker detector is also expected to exhibit dark counts in a narrower relative current interval. Experimentally, however, we only observed an exponential dark-count rate for the 5 nm sample. The absence of dark counts in the case of $d = 10$ nm (see open symbols in Fig. 4.5) is unclear at this point.

The thickness-dependent energy barrier has not only consequences on dark counts and the switching current but also on fluctuation-assisted photon counts. From the considerations above, one would expect a narrower probabilistic detection regime in the case of thicker detectors, in line with the expected narrower current interval for dark counts. However, as seen from the sharper curve of the 5 nm detector in Fig. 4.5, we actually observe the thinner detector to have an exponential count-rate increase within a narrower relative current interval. Further measurements of detectors with different film thickness are needed to probe the theoretically expected dependences and could also prove an optimal detector thickness with minimum dark counts and high-enough photon counts.

4.4 Summary

This chapter evaluated the performance of SNSPDs made from ALD-NbN thin films. At an operating temperature of 4.2 K, the 5 nm thick detector exhibits photon detection in the deterministic regime above $I_b \approx 0.4 I_{sw}$ at $\lambda = 385$ nm and above $\approx 0.6 I_{sw}$ at 658 nm, as observable from the saturating count rates over bias current (in logarithmic scale). Because of the larger cross-section (see Chapter 2), the onset of count-rate saturation of the 10 nm thick sample is shifted to higher relative bias currents. The photon-count rates of both detectors are linearly dependent on the incoming photon rate (CR $\propto$ PR) which corresponds to a constant detection efficiency independent of the optical power and is a strong indication for single-photon detection. Spectral characterizations show a count-rate plateau below $\lambda \approx 1$ µm and a strong decrease in CR when detecting longer-wavelength photons. This is in accordance with the expected bias- and wavelength-dependent minimal photon energy detectable by SNSPDs. Comparing 5 nm and 10 nm thick detectors, the thinner SNSPD exhibits an increase in cut-off wavelength by a factor of 1.6 at the same relative bias current.

Estimations of the thickness-dependent vortex-entry barrier together with the obtained experimental results have shown that, due to the dependence of the energy barrier $dF \propto d$, thicker detectors can be operated closer to the depairing current and potentially feature fewer dark counts. For the same reason of suppressed thermally activated vortices, however, detectors from thicker films are less sensitive to photons of lower energy, which poses a compromise in detector development.

Compared to reported detectors out of sputtered NbN with similar cross-sections and operating temperatures, the ALD-NbN-SNSPDs show slightly longer cut-off wavelengths [85]. Significantly higher λ_c values and saturated count rates up to around 3 µm were demonstrated for sputtered, NbN-based detectors before, albeit at a much lower operating temperature of 1.5 K and for a smaller nanowire cross-section of 5.5 nm × 30 nm [10].

Overall, the results depict ALD-based nanowire detectors as a promising alternative to sputtered detectors, especially in conjunction with the large-area uniformity results from Chapter 3. Nevertheless, while the investigated SNSPDs show saturated single-photon detection up to near-infrared range, further thin-film development is necessary to optimize ALD-NbN for SNSPDs. Besides, operating the detectors below 4.2 K will also be needed to probe the performance limit.

5 Frequency-multiplexed SNSPDs

Exceptional detection properties have been shown by single-pixel SNSPDs. For many practical applications like imaging and spectroscopy, however, arrays of SNSPDs are required. An SNSPD array which can take advantage of the outstanding timing precision and count rates combined with high efficiencies from visible light to mid-IR would enable new fields of applications for SNSPDs.

As outlined in Section 2.5, frequency-domain multiplexing (FDM) is one of the various architectures that have been shown for such arrays [186]. These so-called RF-SNSPDs can be biased individually, read out by a single RF line, and offer spatial resolution of multi-photon events. However, there are also several shortcomings of this multiplexing technique which will be addressed in this chapter. The oscillating bias current is expected to have several consequences on the detector operation. On the one hand, it should lead to a stable operation without the risk of the nanowire latching into the resistive state, which will be the subject of Section 5.2. On the other hand, the microwave bias is also expected to modulate the intrinsic detection efficiency $\eta_{\text{int}}(I_{\text{b}})$ accordingly, leading to a reduced duty cycle of the detector. A detection scheme which takes advantage of the modulated efficiency will be introduced and analyzed in Section 5.3. Another disadvantage is the elaborate readout necessary to differentiate between detection events of different pixels. In Section 5.4, an alternative readout method based on direct digitization and matched filters will be investigated.[1]

5.1 Array design and fabrication

The devices used in this chapter were designed and fabricated by A. Kuzmin and originally intended for experiments with microlenses (see Section 6.1). Here, we use the same layout to analyze the frequency-domain multiplexing scheme instead.

[1] A third shortcoming of RF-SNSPDs, the low fill factor owing to the large resonator, will be addressed in Section 6.1 of the next chapter.

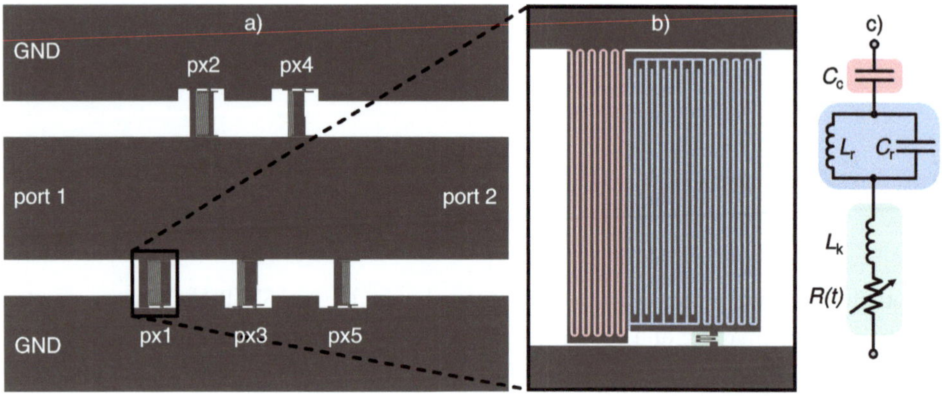

Figure 5.1: Design of a five-pixel RF-SNSPD array. (a) Array consisting of five pixels connected to a common transmission line from port 1 to port 2. (b) Layout of pixel 1 with a coupling capacitor (red), an LC parallel circuit (blue) and a nanowire (green). (c) Equivalent circuit of a single pixel with colors matching the highlighted elements in (b).

Throughout this chapter, an RF-SNSPD array with five pixels in the frequency range (1.9 to 3.5) GHz is used as depicted in Fig. 5.1. The pixels each have a target resonance width of $\Delta f \approx 15\,\mathrm{MHz}$ and are all connected to a common 50 Ω transmission line of width 250 µm and with a gap of 75 µm.

A close-up of pixel 1 is shown in Fig. 5.1b. The resonant circuit (highlighted in blue) consists of a meandered inductor and an interdigitated capacitor which define L_r and C_r, respectively. Both are connected in parallel and are defined by 1 µm-wide structures and gaps. The inductor length is kept constant at 1105 µm for all pixels. With a square kinetic inductance of $L_{k,sq} \approx 45\,\mathrm{pH}$ of the 5 nm thick NbN in use, L_r amounts to about 50 nH. To vary the resonance frequency, parallel capacitance C_r is tuned by modifying the finger number and length across the pixels. This is in contrast to Ref. [188] and [186] where a variable inductance L_r was used to set f_r. There, it was supposed, however, that the contribution of the inductor's turns to L_r lead to insufficiently controllable values of f_r. As a result, the resonance frequencies of the fabricated device were not distributed with equal distance as designed. In an effort to mitigate this problem here, C_r was chosen instead as the parameter to vary f_r.

Each pixel is coupled to the transmission line by an interdigitated capacitor (shaded red in Fig. 5.1b) of 1 µm in both width and gap. In order to keep the resonance depth constant over all pixels, C_c was adjusted in conjunction with C_r with the help of simulations in Sonnet em.

The meandered nanowire (shaded green), which constitutes the active area of each pixel and connects each resonant circuit to the ground plane, has a width of ≈100 nm and a length of ≈15 µm. With 150 squares, $L_k \approx 7\,\text{nH}$ which corresponds to a ratio $L_r/L_k \approx 7$ and ensures that the strong bias dependence of L_k close to the switching current of the nanowire does not significantly influence the resonance frequency (see explanation in Section 2.5.2).

The array is fabricated from a single, 5 nm thick NbN layer on an R-plane sapphire substrate. Similar to the process described in Section 4.1.2, the device is patterned by EBL and Ar-ion milling. Different from previous chapters, detectors characterized in this chapter were fabricated from *sputtered* NbN films. Using ALD-NbN for the detectors would have been possible as well. In this case, the higher resistivity and smaller energy gap of the ALD-NbN films would have resulted a slightly higher kinetic inductance (Eq. 2.4) and therefore in a shift of f_r toward higher frequencies for the same device geometry. This could have been compensated by adjusting the resonator dimensions.

Two nominally identical devices were fabricated, featuring slightly different resonance frequencies. Results obtained from one device are shown in Sections 5.2 and 5.3, whereas results of the other device are presented in Section 5.4 (and later in Section 6.1).

5.2 Self-resetting behavior

This section contains parts from previously published content:
[241] E. Knehr, A. Kuzmin, S. Doerner, et al., "Synchronous single-photon detection with self-resetting, GHz-gated superconducting NbN nanowires", Applied Physics Letters ***117****, 132602 (2020).*
Permission to include adapted text was obtained from the authors.

When a photon is absorbed by a DC-biased nanowire, a normal-conducting domain (hotspot) can form and its resistance manifests itself as a voltage pulse. If the nanowire's kinetic inductance is large enough (the nanowire is sufficiently long), the superconductive state can be restored and the detector resets. Latching into the normal state can occur if the nanowire is too short (see also Section 2.2.3). In this case, a stable negative electrothermal feedback by the load impedance keeps the induced hotspot inside the nanowire in a self-heating state [53]. The hotspot's lifetime strongly depends on the applied bias current [54, 242]. For currents below the retrapping current, the hotspot rapidly collapses. The latching behavior of the detector also depends on the readout circuit. The commonly used AC-coupling of the readout amplifier has several consequences on the detector behavior and can lead to latching at high

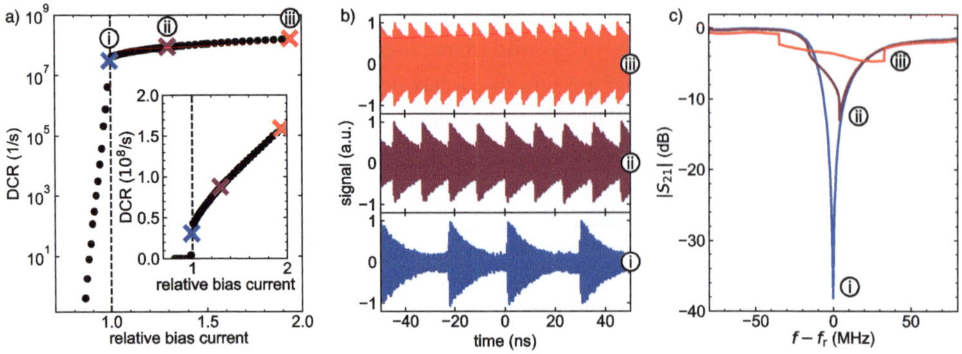

Figure 5.2: Self-resetting behavior of an RF-SNSPD pixel. (a) Dark-count rate over relative bias current in semi-log and linear scale (inset). (b) Recorded waveforms of the detector signal at the bias levels highlighted in (a). (c) Microwave transmission of the RF-SNSPD pixel over frequency $f - f_r$ at bias levels marked in (a).

photon rates when biased below but close to the switching current. This is described in detail in Ref. [107].

For applications, detector latching poses a problem, for example, in laser ranging of space debris where intense photon backscattering at lower altitudes can cause the SNSPD to permanently switch to the resistive state [243]. In order to overcome this problem, Zhang *et al.* showed a DC-SNSPD operation with periodical off-states using a programmed bias source to reset the detector [243]. Another example are blinding attacks in SNSPD-based quantum key distribution (QKD) systems, provoked by bright light pulses. As a possible countermeasure, Elezov *et al.* have tested an auto-reset system for SNSPDs using a voltage-comparator [244].

In the case of an RF-SNSPD, the nanowire is biased by a microwave current I_{rf} at the resonance frequency f_r of the nesting lumped-element resonator, which is capacitively coupled to a 50 Ω feed line. After photon absorption, a growing hotspot damps the resonator and therefore leads to an increased transmission coefficient $|S_{21}|$ on the feed line at f_r, which is used as the detector response.

With this scheme, self-resetting of the nanowire is enforced by the oscillating current and the decay time of the resonator τ_d. The hotspot growth time is limited to few periods of the RF current, since the hotspot collapses at low currents on a timescale of ≈10 ps. Additionally, the damping of the resonant circuit upon hotspot generation causes an RF current redirection out of the nanowire to the feed line for at least several periods. This effectively resets the detector after every detection event.

The self-resetting behavior leads to a stable operation of the RF-SNSPD also at bias amplitudes close to the switching current. This is demonstrated by measuring the bias-dependent dark-count rate of a single pixel of the array (Fig. 5.2a). Below a certain current amplitude (associated with switching current I_{sw} and shown as 1.0), there is an exponential dependence of the DCR on the bias and dark counts are random. At I_{sw}, DCR reaches a value of $\approx 1/\tau_{\mathrm{d}}$. For relative bias currents above 1.0, however, the DCR does not drop to zero but linearly rises and the pulses are periodic. We associate this operation with a bolometric regime in which the RF-SNSPD demonstrates a self-oscillatory behavior, seen as strictly periodic "dark counts" with a rate linearly proportional to the ratio of the bias amplitude to the switching current (DCR $\propto I_{\mathrm{rf}}/I_{\mathrm{sw}}$, as seen on the linear scale in the inset of Fig. 5.2a). If any excess optical power heats the nanowire, I_{sw} is reduced and the apparent DCR increases correspondingly. This way, the RF-SNSPD with DCR $> 1/\tau_{\mathrm{d}}$ operates as a bolometric detector with "power-to-frequency" responsivity, similarly to a superconducting noise bolometer [245]. For the bias levels highlighted in Fig. 5.2a, recorded RF pulses are depicted in Fig. 5.2b. One can see dark-count pulses with varying pulse-to-pulse times in (i), whereas at relative bias currents above 1.0 the pulses become periodic with a pulse rate $\propto I_{\mathrm{b}}$ in (ii) and (iii). In addition, microwave transmission measurements of the RF-SNSPD at the corresponding bias levels are shown in Fig. 5.2c. Curve (i), for which the current amplitude through the nanowire is still below I_{sw}, exhibits a resonance depth close to the value measured at much lower RF power levels. When the power is increased and the current amplitude exceeds I_{sw}, the finite resistance of the nanowire leads to an increased transmission $|S_{21}|$ at f_{r} which is seen as a rapidly vanishing resonance. However, the cratered resonance does not indicate that the nanowire is locked in the normal state but, in fact, periodically switches between superconducting and normal states.

5.2.1 Simulation of DC- and RF-biased detectors

From the previous section, it is clear that the oscillating bias current through the nanowire effectively resets the detector even for I_{rf} with an amplitude above I_{sw}. The investigated RF-SNSPD pixel, however, has a comparatively long reset time (ringdown time of the resonator) of $\approx$20 ns. During normal operation, most DC-biased SNSPDs from common materials do not exhibit latching at this reset time either. In order to substantiate the latching-free behavior of the RF-SNSPD, the operation of an RF-biased nanowire with much shorter ringdown time on the order of 1 ns was simulated using an electrothermal model of the detector, and compared to the behavior of a DC-biased nanowire. To this end, we implemented the hotspot dynamics provided by Kerman/Berggren (see Section 2.2.3 and Refs. [53, 60]) using the Simscape Electrical library of Simulink.

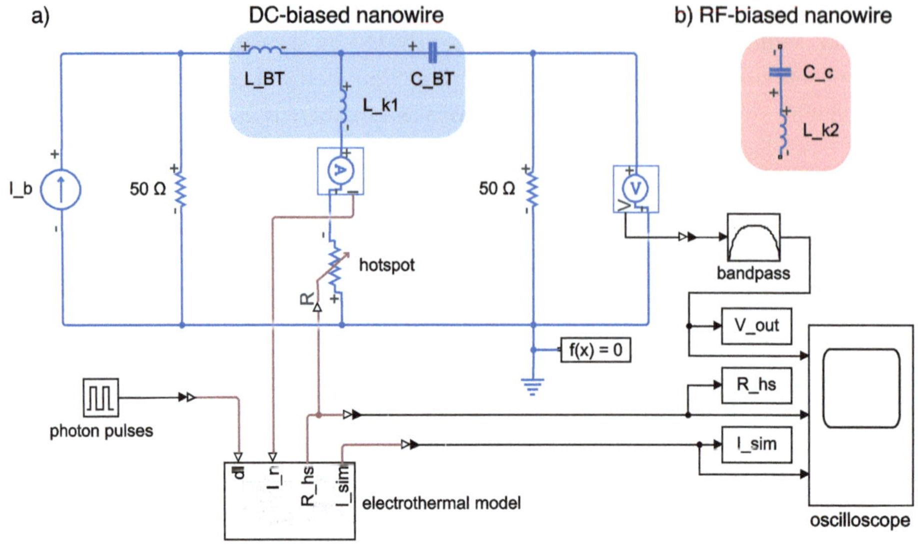

Figure 5.3: Simscape schematic to simulate the latching behavior of DC- and RF-biased nanowires. The DC-biased nanowire was simulated using the schematic in (a). For the RF-biased nanowire, the kinetic inductance and bias tee [blue background in (a)] were substituted by the capacitively coupled inductance in (b).

In Fig. 5.3 the schematic of the model is shown. The hotspot dynamics according to Eqs. 2.6 and 2.7 are implemented by the block labelled as "electrothermal model". It provides the hotspot resistance R_{hs} from nanowire current I_{n} and an additionally injected current $\mathrm{d}I$. The following device parameters are used: $I_{\mathrm{sw}} = 30\,\mu\mathrm{A}$, $R_{\mathrm{sh}} = 350\,\Omega$, $w = 110\,\mathrm{nm}$, $v_0 = 236\,\mathrm{m/s}$, and $\psi = 76$. A photon detection event in the nanowire causes a local suppression of the switching current, leading to a detection event if I_{sw} is reduced below the applied current through the nanowire. Following Ref. [60], we simulate the detection event by injecting an excess current $\mathrm{d}I$ which shortly increases the current through the nanowire beyond its switching current I_{sw}. For this purpose, the block named "photon pulses" emits current pulses of amplitude $10\,\mu\mathrm{A}$.

Two different configurations, DC and RF bias, were simulated. For the DC-biased nanowire (Fig. 5.3a), the kinetic inductance was set to $L_{\mathrm{k1}} = 40\,\mathrm{nH}$, resulting in a fall time of $\approx 0.8\,\mathrm{ns}$. It was biased by a constant current $I_{\mathrm{b}} = 0.95 I_{\mathrm{sw}}$ through the inductive port of a bias tee ($L_{\mathrm{BT}} = 1\,\mu\mathrm{H}$) and read out through capacitance $C_{\mathrm{BT}} = 4\,\mathrm{nF}$. The RF-biased nanowire (Fig. 5.3b) with the same switching current was modeled by an LC resonator consisting of the nanowire's kinetic inductance $L_{\mathrm{k2}} = 5\,\mathrm{nH}$ and capacitance $C_{\mathrm{c}} = 1.4\,\mathrm{pF}$. With $f_{\mathrm{r}} = 1/(2\pi\sqrt{LC})$, the device exhibits a resonance at $\approx 1.9\,\mathrm{GHz}$ and a simulated resonance width of $\approx 860\,\mathrm{MHz}$. The latter corresponds to a quality factor of 2.2 and a ringdown time of 0.4 ns, about half the value of the DC-biased detector. The nanowire was biased by an alternating current with an amplitude close to the switching current of the nanowire. Note that this configuration differs from the RF-SNSPD circuit used in the other sections of this chapter (see schematic in Fig. 5.1c). Here, the nanowire's kinetic inductance is directly used as the inductive part of a series *LC* resonator and the separate, parallel *LC* circuit is omitted.

In both configurations (DC and RF bias), dependence $L_k(I_b)$ was neglected. Typically, the maximum variation of the inductance is ≤20 % over the bias range (see Ref. [60], or Fig. 7.4 in Section 7.3) and therefore does not change the time scales of the simulated DC-biased nanowire behavior significantly. For the RF-bias case, $L_k(I_b)$ can, in general, prohibit the detector operation when $f_r(L_k)$ leads to a large enough resonance shift such that the bias power at a specific frequency does not couple into the resonant circuit anymore after a detection event. However, due to the large resonance width Δf of this particular circuit (larger than the expected shift in f_r), the $L_k(I_b)$ dependence is not expected to hinder the detector operation and neglecting the dependence is therefore adequate.

Depending on the configuration, the pulse period of dI slightly varied: in the case of DC bias, the period was set to $\tau = 5$ ns, while for the RF bias $\tau = 10/f_r$. The latter was chosen to avoid problems with the reduced duty cycle of the nanowire with an oscillating bias current. This will be investigated in Section 5.3.3. By defining the photon pulse period by means of the resonance frequency, the phase of the bias current at the time of the photon event is kept constant. If the phase is chosen such that dI is injected at bias currents close to the maximum, it is ensured that every photon event leads to a growing hotspot.

Figure 5.4 shows simulated time traces over 100 ns of the detector signal[2], the hotspot resistance R_{hs}, and current $I_{sim} = |I_n| + dI$ for both DC (left column) and RF bias (right column). Photon pulses are injected after a delay of 10 ns to account for the settling time of both circuits. These pulses are seen as short peaks of $\Delta I = 10\,\mu$A in the bias traces.

The DC-biased nanowire exhibits periodically emerging hotspots as expected, with $R_{hs} \approx 850\,\Omega$ and a relaxation time of ≈230 ps. Over the load resistance of 50 Ω, this is seen as voltage pulses of ≈1.2 mV amplitude and ≈1 ns decay time. About 2 ns after every detection, the bias current through the nanowire has restored. Notably, the voltage signal depicted in Fig. 5.4a returns to a value below zero after every detection event, equivalent to a negative load current. This is the result of the AC-coupled readout, namely the discharge of the capacitor used to readout the detector (C_{BT} in Fig. 5.3), and is also known as "back-biasing" [60]. The effect was

2 To approximate a real-world setup, the voltage across the load resistance of 50 Ω was filtered by a second-order Butterworth band-pass filter with edge frequencies of 15 MHz and 8 GHz.

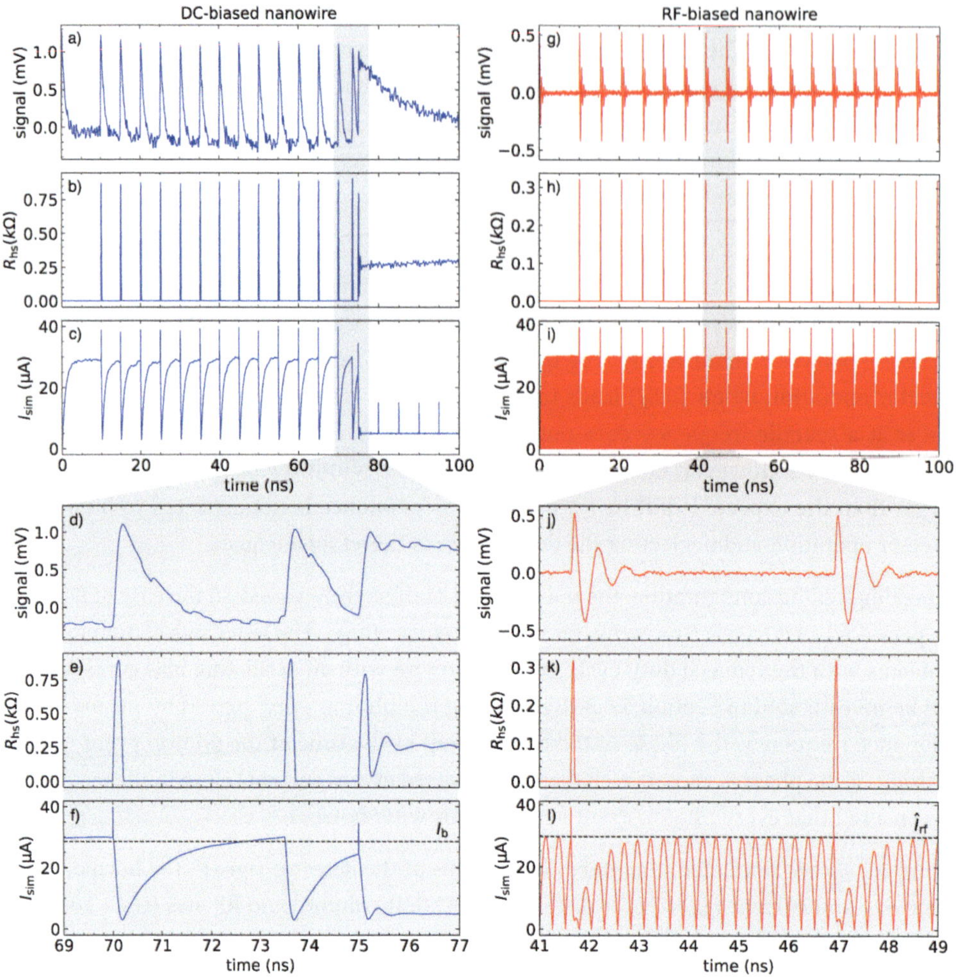

Figure 5.4: Simulation of the detector signal, hotspot resistance R_{hs}, and current $I_{sim} = |I_n| + dI$ for DC bias (left column) and RF bias (right column) using the models from Fig. 5.3. The plots in the lower half are zooms of the respective time spans denoted in the graphs above. In (f) and (l), dashed lines indicate bias current I_b and bias amplitude $\hat{I}_{rf}$, set by the respective bias source.

previously described by Kerman *et al.* (compare Fig. 1f in Ref. [107]), arises at high count rates, and also causes a higher current through the nanowire $I_n > I_b$ than what is set by the current source. Indeed, in Fig. 5.4f, a higher current I_n through the nanowire than $I_b = 28.5\,\mu A$ can be discerned before the event at $t = 70$ ns. Eventually, the back-biasing of the nanowire can lead to relaxation oscillations and latching into a stable resistive state. For the particular operation parameters here, only a single oscillation at $t \approx 73.5$ ns between two consecutive photon pulses is observed, before the detector latches after the following photon pulse at $t = 75$ ns. This manifests itself as a steadily growing hotspot resistance, which results in a small constant current through the nanowire (disregarding the periodic dI pulses) and a permanent voltage across the load resistance. The particular time at which the nanowire oscillates and latches depends on the applied bias current I_b, the kinetic inductance L_k, the readout capacitance (here only C_{BT}), as well as the readout resistance R_l. For a larger kinetic inductance, we observed multiple relaxation oscillations with decreasing period before the detector latches, comparable to Fig. E2 in the supplementary of Ref. [60]. Similarly, a bigger readout capacitance also prolongs the time span over which oscillations occur before the detector latches. Conversely, the detector operation is more prone to latching for bias currents closer to I_{sw} or a higher load resistance (see, e.g., Fig. 7 in Ref. [60]).

Time traces over the same span for the RF-biased nanowire are shown in Figs. 5.4g–l. The respective detector pulses are slightly shifted with respect to the DC bias because of the different photon-pulse period (see above). From Figs. 5.4k and 5.4l, it is visible that the maximum hotspot resistance at approximately 300 Ω is smaller than the one simulated for the DC-biased nanowire, and drops rapidly as the amplitude of the RF current through the nanowire decreases. This is both due to the oscillation phase and the fast redirection of the RF current to the 50 Ω load resistance. It is also seen that, for the device parameters in use, the hotspot only persists over half a period of the RF current. This behavior depends on the quality factor of the resonator. For a larger quality factor, it was found that the hotspot can persist over few bias oscillations as well. Following a detection event, the bias current through the nanowire is restored after around three full oscillations of I_{rf} (see Fig. 5.4l). Compared to the DC-biased nanowire, the operation is not locked into the resistive state of the nanowire at any time and the detector is effectively reset after every detection event, enabling a stable operation.

It should be noted that the detector signals were not amplified in these simulations. Needless to say, a capacitively coupled amplifier would add a capacitor in series to the load also for the RF-biased nanowire, even in the absence of a bias tee. Nevertheless, even if the amplitude of $|I_{rf}|$ would exceed the switching current of the nanowire, the resulting oscillatory behavior would not lead to the instability of the device, i.e., latching (see Fig. 5.2).

5.3 Synchronized gated detection

This section contains parts from previously published content:
[241] E. Knehr, A. Kuzmin, S. Doerner, et al., "Synchronous single-photon detection with self-resetting, GHz-gated superconducting NbN nanowires", Applied Physics Letters ***117****, 132602 (2020).*
Permission to include adapted text was obtained from the authors.

The intrinsic detection efficiency of an SNSPD depends on the applied bias current. DC-biased SNSPDs are biased by a constant current I_b close to I_sw. If the PCR(I_b) curve is saturated in this bias regime, one can assume $\eta_\mathrm{int} \approx 1$. This free-running operation of SNSPDs is an advantage over competing types of single-photon detectors like single-photon avalanche diodes (SPADs), which require a gated operation to mitigate afterpulsing [246].[3] The only detrimental effect on η_int during the normal SNSPD operation is due to the finite current recovery time to the nanowire after a detection event, and becomes relevant only at high count rates. This dependence was investigated in Ref. [52] (see also SDE(CR) measurements in Fig. 4d of Ref. [106]).

On the contrary, RF-SNSPDs are operated by an oscillating bias current I_rf only. Because of dependence $\eta_\mathrm{int}[I_\mathrm{rf}(t)]$, the RF-SNSPD is not expected to be active all the time, i.e., it is in gated mode. In the case of random photon arrivals, this should lead to a reduced detection efficiency compared to a similar DC-biased SNSPD. For known photon-arrival gates, on the other hand, the detection efficiency can be optimized by synchronizing the RF-SNSPD to the base clock of the optical signal. As a result, the RF-SNSPD's detection efficiency should become comparable to a DC-biased SNSPD. In Ref. [189], the direct comparison of a DC- and RF-biased SNSPD has demonstrated that a similar asynchronous η_int can be achieved. However, the comparison requires an accurate determination of the RF current through the nanowire which is difficult. Moreover, η_int of an RF-SNSPD synchronized to a modulated light source at half of the gating frequency has revealed a sensitivity to the relative phase between the RF bias and optical modulation, although the change in η_int was not well understood. In this section, results of an RF-SNSPD fully synchronized to a modulated light source in GHz range are presented. Based on these results, the influence of the sigmoidal $\eta_\mathrm{int}(I_\mathrm{b})$ dependence on the duty cycle is estimated in the end.

[3] For different purposes, a gated or quasi-gated approach for SNSPDs has been adopted before. Besides the aim to mitigate the latching problem (see previous section), Akhlaghi and Majedi presented a gated mode at 625 MHz with a DC offset to increase the maximum count rate [247].

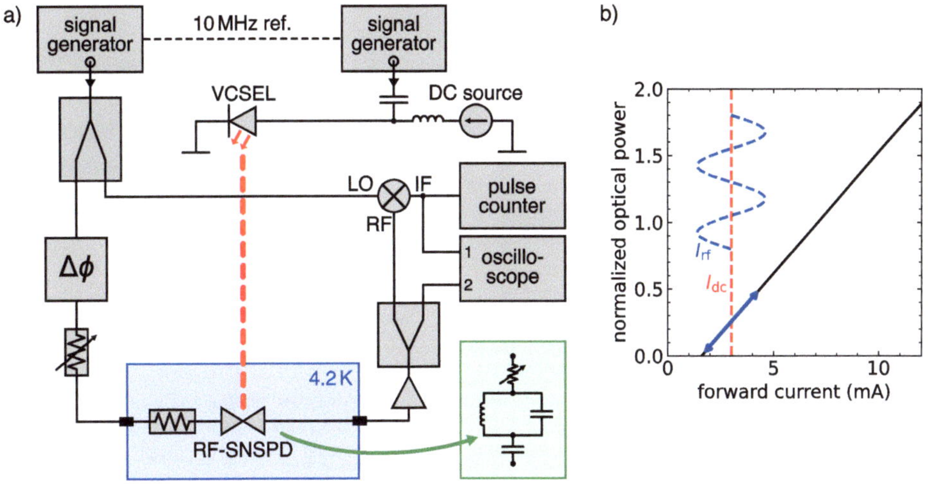

Figure 5.5: (a) Schematic of the experimental setup for synchronized photon detection. The RF-SNSPD (lumped-element model in green) is biased by a single tone from a signal generator and its signal is read out both directly as well as down-mixed. The VCSEL is biased by a second, phase-locked signal generator with an RF current offset by a DC source. (b) Sketch of the VCSEL operation with DC (red) and RF bias (blue). The black line is loosely based on the output power vs. forward current graph provided in the data sheet.

For the following investigations, pixel 1 from a five-pixel array (see Fig. 5.1) is used and operated at its resonance frequency $f_r = 1.9$ GHz. Thus, the gating frequency is $f_g = 2f_r = 3.8$ GHz.

5.3.1 Experimental setup

The measurement setup for the synchronized photon detection is depicted in Fig. 5.5a. As a GHz-modulated light source, we use a vertical-cavity surface-emitting laser diode (VCSEL, Optek Technology OPV314YAT) emitting light at a wavelength of 850 nm and allowing for modulation frequencies $f_{opt} \lesssim 4.5$ GHz.

Two phase-locked RF signal generators provide the RF bias currents for both the laser diode and the RF-SNSPD. The laser diode is biased by a DC current of 3 mA and an RF current with an amplitude large enough to turn it off at the minimum, resulting in an optical output power modulated in GHz range similar to on-off keying (OOK). A visualization of this is shown in the output power vs. forward bias sketch of the VCSEL operation in Fig. 5.5b. A fast photodiode (Thorlabs PDA8GS, DC–9.5 GHz) was used to accurately measure the spectrum of the optical

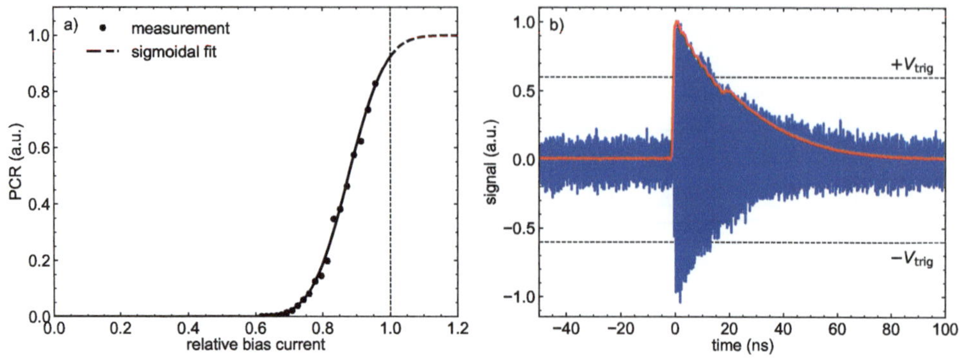

Figure 5.6: Current-dependent photon-count rate and detector pulse. (a) Measured photon-count rate dependent on the relative bias current. The solid line is a fit by Eq. 2.10. (b) Single RF (blue) and downmixed (red envelope) pulse upon photon absorption. The trigger levels indicated by dashed lines are used to record the graph in Fig. 5.7.

modulation of the VCSEL. After setting the appropriate bias level, the VCSEL output was optically coupled to the detector using a multi-mode fiber (Thorlabs FG105LCA).

The RF-SNSPD is biased by an RF current only. The output power of the signal generator is set to 21 dBm, such that, combined with the loss of −6 dB from the splitter, the mixer is fed with a power within the specified LO (local oscillator) drive level range of (13 to 18) dBm. The exact bias power level of the RF-SNSPD pixel is then set by a combination of a variable attenuator for coarse adjustment and the signal generator for fine-tuning (within ±1 dBm). The pulses from the detector are amplified by 37 dB using a room-temperature amplifier and read out by a 32 GHz real-time oscilloscope (port 2). Additionally, the RF pulses are downmixed by an analog mixer using the RF port as input, the IF (intermediate frequency) port as output and the bias signal as LO (homodyne mixing scheme). The resulting envelope is read out with the oscilloscope (port 1) and an ordinary pulse counter to measure count rates.

5.3.2 Phase-sensitive detection efficiency

The current-dependent photon-count rate (PCR) of the detector was measured at a constant VCSEL output power ($I_{\mathrm{dc}} = 3\,\mathrm{mA}$) and is plotted in Fig. 5.6a. As it is seen, the curve does not saturate up to a current at which the DCR becomes dominant.

In Fig. 5.6b, a single detector response upon photon absorption and its envelope are shown. The decay time $\tau_{\mathrm{d}} = Q/(\pi f_{\mathrm{r}}) \approx 20\,\mathrm{ns}$ is determined by the loaded quality factor Q of the resonant circuit and is consistent with the target resonance width of 15 MHz ($Q \approx 120$).

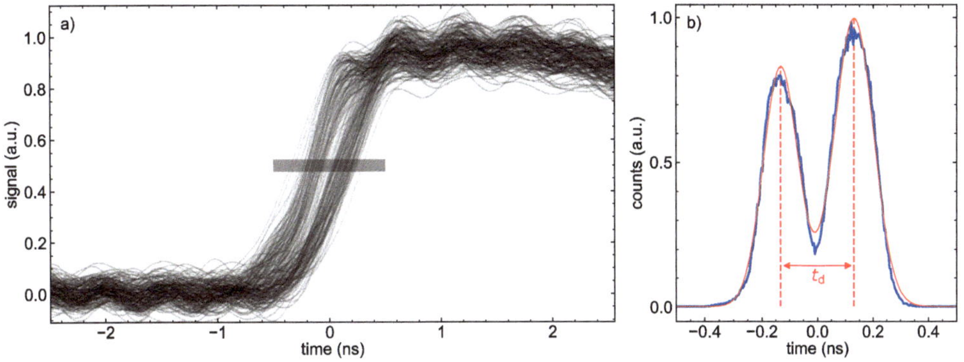

Figure 5.7: Time traces and histogram of pulse envelopes. (a) Rising edge of multiple pulse envelopes measured by triggering on the RF pulses at the trigger levels indicated in Fig. 5.6b. (b) Normalized histogram of the number of counts over time at the rising edge of the pulses [indicated by the semitransparent rectangle in (a)] with measurements in blue and Gaussian fits in red. Time difference t_d is taken as the difference between the mean values of both fits.

Figure 5.7a shows multiple traces of the envelope (channel 1 of the oscilloscope, Fig. 5.5), which are obtained with constant optical power by triggering on the original RF pulse (channel 2) exceeding a defined range ($-V_\mathrm{trig}$ to $+V_\mathrm{trig}$, see Fig. 5.6b). The polarity and amplitude of the first oscillation of the RF pulse depends on the phase of the bias current at the time of photon absorption. With this trigger method, one can visualize the distribution of count events relative to the phase of the bias. Two distinct groups of envelopes can be observed. This is also seen from the histogram of rising edges which is depicted in Fig. 5.7b. The time difference between the two peaks amounts to $t_\mathrm{d} \approx 263\,\mathrm{ps}$, which corresponds to a frequency $f \approx 3.8\,\mathrm{GHz}$ or twice the resonance frequency.

The grouping of pulses shown in Fig. 5.7a can be explained by the nonsaturated count rate characteristic CR (I_b) of the NbN detector at $\lambda = 850\,\mathrm{nm}$ and $T = 4.2\,\mathrm{K}$ (Fig. 5.6a). Since, in this case, η_int does not reach 100%, the majority of the detection events will originate from bias points near the extrema ($\pm I_\mathrm{rf,max}$). The measured grouping of pulses around these two bias points of opposite polarity implies that the detector is active twice in a period of the RF bias. The active time span (gate duration) can be estimated by the time during which $|I_\mathrm{rf}| \geq I_\mathrm{det}$, with the minimal detection current I_det for a particular photon energy.

In order to show that the RF-SNSPD's detection efficiency might be optimized by synchronizing the detector to the optical signal, the phase-sensitive changes of the count rate are investigated. The measurements are conducted with the light modulated either with the resonance frequency of the RF-SNSPD ($f_\mathrm{opt}/f_\mathrm{r} = 1$) or with its gating frequency ($f_\mathrm{opt}/f_\mathrm{r} = 2$). To slowly sweep the relative phase between the RF bias and the optical signal, the frequency

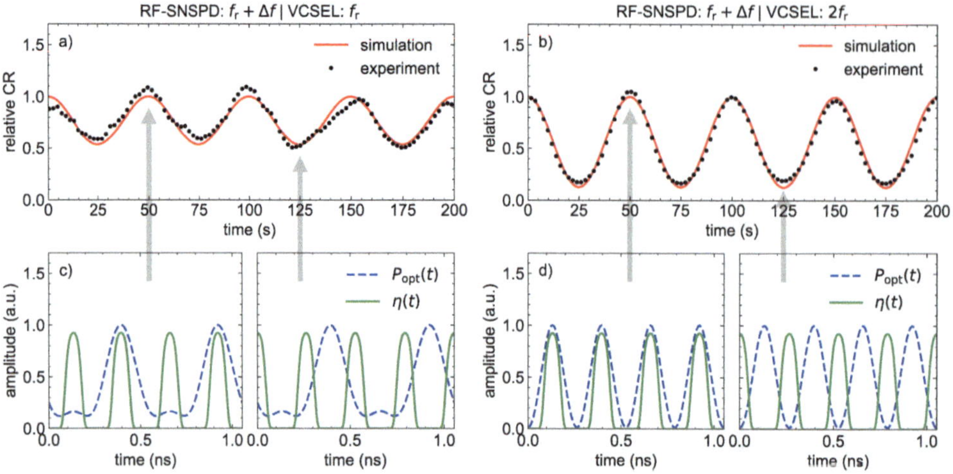

Figure 5.8: Measurement and simulation of modulated count rate CR(t) for a synchronized detection and beat frequency $\delta f = 10\,\text{mHz}$. (a), (b) Experimental count rates CR(t) for $f_{opt}/f_r = 1$ and $f_{opt}/f_r = 2$ compared to simulated curves obtained from Eq. 5.4. (c), (d) Simulation of optical signal $P_{opt}(t)$ and instantaneous detection efficiency $\eta(t)$ for both modulation frequencies f_{opt}. In each case, full alignment and misalignment between $P_{opt}(t)$ and $\eta(t)$ are depicted [in (d) corresponding to phases 0° and 180° of $\eta(t)$ relative to $P_{opt}(t)$].

of the RF bias is detuned with respect to the modulation frequency of the laser diode by $\delta f = 10\,\text{mHz}$. This small detuning ($\delta f / f_r \approx 10^{-12}$) should lead to a slow periodic modulation of the count rate with the beat frequency.

The phase-sensitive detection efficiency was analyzed experimentally (Figs. 5.8a and 5.8b), and oscillations of the count rate with a period of 50 s are indeed visible. The measured count rate modulation index $m = (\text{CR}_{max} - \text{CR}_{min})/(\text{CR}_{max} + \text{CR}_{min})$ amounts to $m \approx 0.71$ for $f_{opt}/f_r = 2$. For the case $f_{opt}/f_r = 1$, the modulation index is smaller ($m \approx 0.30$) and depends on the modulation amplitude of the laser diode. The modulation period of 50 s observed for a detuning $\delta f = 10\,\text{mHz}$ corresponds to a detuning $\delta f_g = 20\,\text{mHz}$ in gating frequency, which is in accordance with the detector being active twice each bias period (Fig. 5.7).

To describe the measurement results, we simulate the CR(t) dependence using the instantaneous detection efficiency $\eta(t)$ obtained from the fit of CR(I_b). For this, we take a sinusoidal RF bias current

$$I_{rf}(t) = \sin\left[(\omega_r + \delta\omega)\, t\right], \tag{5.1}$$

with $\omega_r = 2\pi f_r = 2\pi \times 1.9\,\mathrm{GHz}$ and $\delta\omega = 2\pi\delta f = 2\pi \times 10\,\mathrm{mHz}$. The instantaneous detection efficiency $\eta(t)$ is then obtained by fitting

$$\eta(t) = a\left[1 + \mathrm{erf}\left(b\,|I_{\mathrm{rf}}(t)| - c\right)\right] \tag{5.2}$$

to the experimental data [Fig. 5.6a, assuming $I_b = I_{\mathrm{rf}}(t)$] with fitting parameters a, b, and c (equivalent to Eq. 2.10). Since $\eta(t)$ in Eq. (5.2) depends on the modulus of the RF bias $|I_{\mathrm{rf}}| \propto \sqrt{1 - \cos(2\omega_r t)}$, it oscillates with the frequency $f_g = 2f_r$.

The amplitude-modulated optical signal is modeled as

$$P_{\mathrm{opt}}(t) = a_0 + a_1\cos\left(\omega_{\mathrm{opt}}t\right) + a_2\cos\left(2\omega_{\mathrm{opt}}t\right), \tag{5.3}$$

with the modulation frequency of the laser diode $\omega_{\mathrm{opt}} = 2\pi f_{\mathrm{opt}}$ and experimentally measured spectral coefficients a_0, a_1, and a_2 for our particular laser diode at the used operation point and both frequencies f_{opt}. Both the instantaneous detection efficiency $\eta(t)$ and the normalized optical power $P_{\mathrm{opt}}(t)$ are plotted over time in Figs. 5.8c and 5.8d for operation frequencies $f_{\mathrm{opt}}/f_r = 1$ and $f_{\mathrm{opt}}/f_r = 2$, respectively.

The time-dependent count rate is estimated using Mandel's formula (Eq. 8 in Ref. [248]) by

$$\mathrm{CR}(t) \propto \int_t^{t+\tau} \eta(t') \cdot P_{\mathrm{opt}}(t')\mathrm{d}t' \tag{5.4}$$

for the mean number of photon clicks in time interval τ.

A comparison of the simulated and measured count rate $\mathrm{CR}(t)$ is shown in Figs. 5.8a and 5.8b. As indicated by the arrows, the maxima and minima of $\mathrm{CR}(t)$ correspond to detector gates fully aligned and misaligned to the optical pulses, depicted in Figs. 5.8c and 5.8d.

For $f_{\mathrm{opt}}/f_r = 2$, in which case the detector and the laser diode are operated with the same frequency and phase, the measurement data coincide well with the simulated curve in both beat frequency and modulation index m. In this case, the amplitude a_2 of the second harmonic of $P_{\mathrm{opt}}(t)$ was found to be very small. The large modulation index signifies that the detection efficiency can be maximized by synchronizing the detector to the light source. The attenuation of stray light and suppression of dark counts can be quantified by the factor $(1 + m)\cdot(1 - m)^{-1}$, with a value of about 6.0 for $f_{\mathrm{opt}}/f_r = 2$.

For $f_{\mathrm{opt}}/f_r = 1$, the modulation of $\mathrm{CR}(t)$ is only due to the second harmonic of $P_{\mathrm{opt}}(t)$, which has a significant amplitude a_2 originating from the nonlinear current-voltage characteristic of the laser diode. The latter also explains the observed dependence of modulation index m on the bias amplitude of the diode. If a_2 was negligible, the shared area under the curves $P_{\mathrm{opt}}(t)$

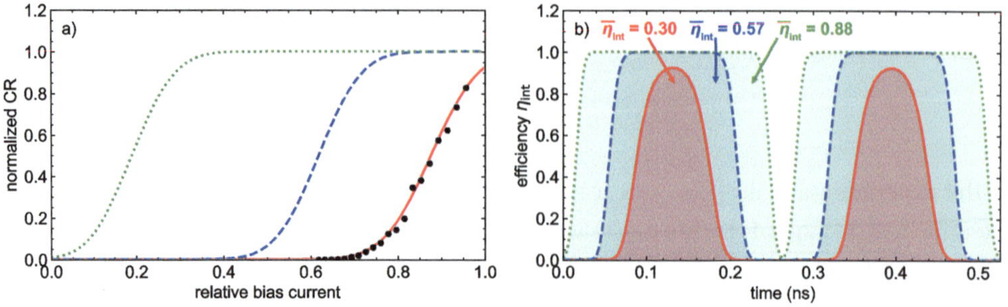

Figure 5.9: Simulation of different CR(I_b) curves (a) and corresponding instantaneous detection efficiencies $\eta_{int}(t)$ according to Eq. 5.2 over one period of f_r (b). The black dots in (a) indicate measurement results. Average detection efficiencies $\overline{\eta}_{int}$ in (b) are calculated from the integral of $\eta_{int}(t)$.

and $\eta(t)$ in Fig. 5.8c would be constant over time and, thus, no modulation of CR(t) would be observed for $f_{opt}/f_r = 1$.

All measurements in this section were conducted using a single pixel. A gated operation of a whole array of pixels with different resonance frequencies is only possible within limits. Assuming fixed frequencies $f_{r,i}$ for pixels $i = 1, \ldots, n$ of the array, the highest operation frequency of the detection system is double the fundamental frequency, i.e., the greatest common divisor (GCD) of all $f_{r,i}$. However, the operating frequency of each pixel could be slightly adjusted around its f_r by few MHz (depending on the resonance width) and the GCD of all gating frequencies thus optimized.

5.3.3 Duty cycle

According to our simple model, the duty cycle of the gated detector is highly dependent on the CR(I_b) characteristic. With a nonsaturated CR(I_b) curve, the investigated detector exhibits a gated detection with photon detection mainly occurring for a current I_{rf} near the extrema $\pm I_{rf,max}$ when $|I_{rf}| \geq I_{det}$, as discussed before (see solid, red curves in Fig. 5.9). In the case of a pronounced count-rate plateau, however, the active time spans can be much longer, which becomes important for detecting photons with random arrival times. This is simulated for two saturated CR(I_b) curves (see dashed and dotted curves in Fig. 5.9). For a broader saturation in Fig. 5.9a, the gate duration increases accordingly, as seen by the instantaneous detection efficiencies $\eta_{int}(t)$ in Fig. 5.9b. The average detection efficiency $\overline{\eta}_{int}$ can be estimated from the integral of $\eta_{int}(t)$, which results in values of 0.30, 0.57, and 0.88 compared to a DC-biased SNSPD constantly operated at $\overline{\eta}_{int} = 1$. For the broadest simulated CR(I_b) curve, the pure RF bias leads to an only slightly decreased efficiency for detecting random photon arrivals and, therefore, becomes comparable to free-running mode. Thus, the duty cycle of an RF-SNSPD with a pronounced count rate saturation can be tuned between gated and quasi-free-running mode by choosing an appropriate RF bias amplitude.

5.4 Matched-filter array readout

In the previous sections, a single pixel of an RF-SNSPD array was operated by a single-tone probe signal from a signal generator. The same microwave signal was also used to downmix the RF pulses coming from the detector, essentially resulting in pulses equivalent to those of conventional DC-biased SNSPDs (see Fig. 5.6b).

To operate a whole array of RF-SNSPD pixels, a frequency comb containing each of the pixel's resonance frequency is necessary. In Ref. [186], an arbitrary waveform generator (AWG) was used to generate the bias signal containing the tone of every pixel. However, the detector signal was only analyzed in frequency-domain by continuously transforming individual time spans of the digitized detector signal using a fast Fourier transform (FFT) and comparing the result to threshold amplitudes for each pixel. The precise timing information of the photon event could be extracted from the RF pulse itself or from the phase spectrum of the complex FFT.

An alternative readout method is explored in this section by using correlation filters instead. In the frequency-domain multiplexing scheme, each pixel operates at a unique frequency. Therefore, templates of each pixel could be used as matched filters to detect the presence of respective detection events in the signal on the common transmission line. Compared to the FFT method used before, the matched filters are applied in time domain and could increase the signal-to-noise ratio (SNR). Matched filters [249] are commonly used to detect known templates in an acquired, noisy signal, for example to detect radar [250] or gravitational-wave signals [251].

5.4.1 Concept

The matched filter is applied by cross-correlating a time trace with a pre-recorded template of the respective pixel. For a discrete trace s and template t_p of pixel p, this is expressed by

$$(t_\mathrm{p} \star s)[n] = \sum_{k=-\infty}^{\infty} t_\mathrm{p}^*[k]\, s[k+n], \tag{5.5}$$

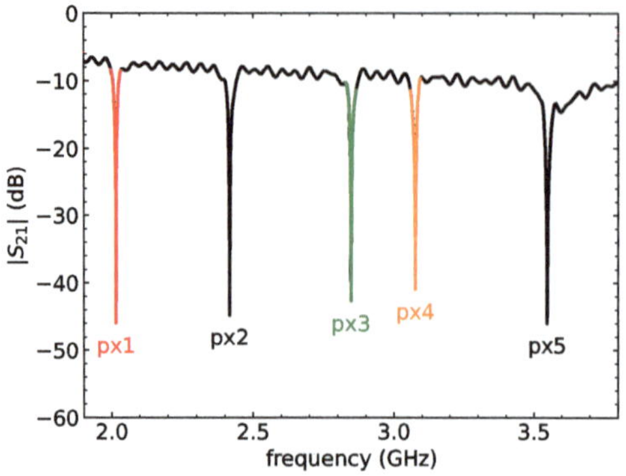

Figure 5.10: Microwave transmission $|S_{21}|$ over frequency of the RF-SNSPD array. Resonances corresponding to all five pixels between 2 GHz and 4 GHz are visible. For all further measurements, the highlighted pixels 1, 3, and 4 were used.

where $t_{\mathrm{p}}^{*}[k]$ denotes the complex conjugate of $t_{\mathrm{p}}[k]$. For finite-length signals $t_{\mathrm{p}}[i], 0 \leq i \leq K$, $s[i], 0 \leq i \leq M$, the boundaries of the summation are defined as $k = \max(0, -n)$ to $\min(K, M - n)$. From the convolution theorem, this is equivalent to $\mathcal{F}^{-1}\left\{\mathcal{F}^{*}(t_{\mathrm{p}}) \cdot \mathcal{F}(s)\right\}$, where $\mathcal{F}(x)$ is the Fourier transform of x. Because the matched filter weights the frequency components of the signal according to the spectrum of the template, the influence of stochastic noise on the signal is reduced and the SNR maximized (in the case of Gaussian noise). The result of cross-correlating the detector signal with a pulse templates are separate traces of detector pulses for each pixel.

5.4.2 Proof-of-principle experiment

A proof-of-principle of the matched filter readout is shown using the five-pixel RF-SNSPD array described in Section 5.1 (see layout in Fig. 5.1). Two pixels of the array are biased simultaneously and the detector signal is directly digitized. The correlation with pulse templates is then carried out by post-processing the recorded signal.

Figure 5.10 depicts the microwave transmission $|S_{21}|$ over frequency recorded with a vector network analyzer (VNA). All five pixels exhibit defined resonances in the range (2.0 to 3.6) GHz with a spacing of ≈400 MHz, except for pixel 4 which is shifted to a slightly smaller frequency than designed. For the following measurements, pixels 1 and 3 were biased, while a third pixel (pixel 4) was not biased and only used to check for false-positive detection-event attributions. The pixels exhibit resonance depths of $\left|S_{21,\mathrm{min,px1}}\right| = -38\,\mathrm{dB}$ ($f_{\mathrm{r,px1}} = 2.01\,\mathrm{GHz}$), $\left|S_{21,\mathrm{min,px3}}\right| = -33\,\mathrm{dB}$ ($f_{\mathrm{r,px3}} = 2.85\,\mathrm{GHz}$), and $\left|S_{21,\mathrm{min,px4}}\right| = -31\,\mathrm{dB}$ ($f_{\mathrm{r,px4}} = 3.08\,\mathrm{GHz}$),

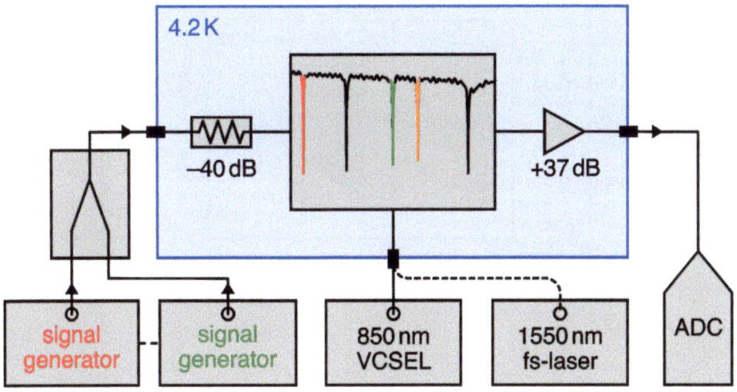

Figure 5.11: Measurement setup to test the matched-filter readout. Two signal generators provide the harmonic signals (tones) with frequencies matching the resonance frequency of the respective pixels. Both bias tones are combined using a signal combiner and coupled into the feedline. The amplified signal on the feedline is directly recorded and digitized by a real-time oscilloscope. Two light sources were utilized: a continuous-wave VCSEL diode at 850 nm wavelength and a pulsed laser at 1550 nm with 100 MHz repetition frequency.

respectively. The resonance depth directly translates to the transmitted power upon photon absorption and therefore the pulse height of a detector click. For each pixel, the loaded bandwidth Δf at -3 dB was $\approx$20 MHz. Therefore, the decay time of a single pulse resulted in $\tau_d = (\pi \Delta f)^{-1} \approx 16$ ns.

To bias and readout the detector array, we use the setup shown in Fig. 5.11. The two pixels operating at $f_r = 2.01$ GHz and 2.85 GHz were biased simultaneously through a common feed line by two separate signal generators and a wideband power combiner. The bias signal was attenuated by 40 dB, which helps to reduce noise on the bias line. As light sources, we used a pulsed laser at 1550 nm wavelength with 100 MHz repetition frequency to obtain the individual detection events in Fig. 5.12, and a VCSEL diode at 850 nm wavelength in continuous-wave mode for photon statistics (Fig. 5.13). The signal at the end of the feed line was amplified by 37 dB by a cryogenic amplifier and directly recorded using a real-time oscilloscope at 10 GSa/s sampling rate.

Before the simultaneous operation, we obtained pulse templates of every pixel by biasing only one pixel at a time and record single-shot pulses. After template acquisition, pixels 1 and 3 were simultaneously biased and illuminated. Because of the large fiber-to-detector distance of 40 mm and the small pitch of 200 µm between both pixels, a homogeneous illumination is assumed, however not strictly necessary for the purpose of the proof-of-principle demonstration.

Waveforms of three recorded events are depicted in the upper row of Fig. 5.12. The increased transmission upon photon absorption is seen as a transient oscillation (RF pulse). These

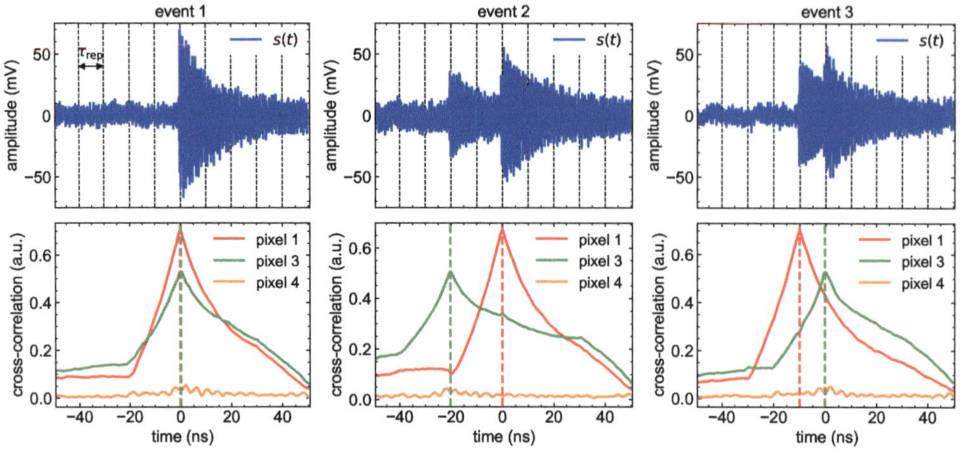

Figure 5.12: Three different recorded waveforms with pixels 1 and 3 triggering at the same time (event 1), pixel 3 triggering two laser repetition periods before pixel 1 (event 2), and pixel 1 triggering one laser repetition period before pixel 3 (event 3). The dashed lines in the upper row denote the timing of the laser pulses with a repetition period $\tau_{\text{rep}} = 10$ ns. The lower row shows upper envelopes of the normalized cross-correlations of the respective event with templates of three different RF-SNSPD pixels. Pixel 4 was not biased during acquisition of the waveforms.

waveforms $s[i]$ were then correlated with templates $t_{\text{p}}[i]$ of each pixel according to Eq. 5.5. The normalized upper envelopes of the results are shown in the bottom row of Fig. 5.12. Each pulse can be assigned to a specific pixel. The events depict two overlapping pulses from pixels 1 and 3 ("event 1"), pixel 1 triggering two laser periods after pixel 3 ("event 2"), and pixel 1 triggering one laser period before pixel 3 ("event 3"). As it can be seen, the matched-filter readout is suited for the attribution of the RF signal to individual pixels, also in the case of two overlapping pulses as shown in "event 1". Depending on the normalization of the cross-correlation, also the amplitude of the RF pulses are replicated by the matched-filter output.

To check for false positives, we used a pre-recorded template of an additional pixel at $f_{\text{r}} = 3.06\,\text{GHz}$ ("pixel 4"), which was not biased during the time of measurement. Pixel 4 is separated by 230 MHz from pixel 3. Only a slight influence of the events attributed to pixel 3 is seen on the matched-filter output of pixel 4.

The minimum achievable frequency separation between two adjacent pixels while maintaining distinguishable matched-filter outputs was tested by simulations with artificial pulse templates at $f_{\text{r,sim}} = f_{\text{r,px3}} + (0 \text{ to } 230)\,\text{MHz}$. For this purpose, a sinusoidal signal of the respective frequency was modulated by a simulated DC pulse with $\tau_{\text{d}} = 20$ ns. At frequencies $f_{\text{r,sim}} = f_{\text{r,px3}}$ and $f_{\text{r,sim}} = f_{\text{r,px4}}$, the matched-filter output closely resembles the one obtained using

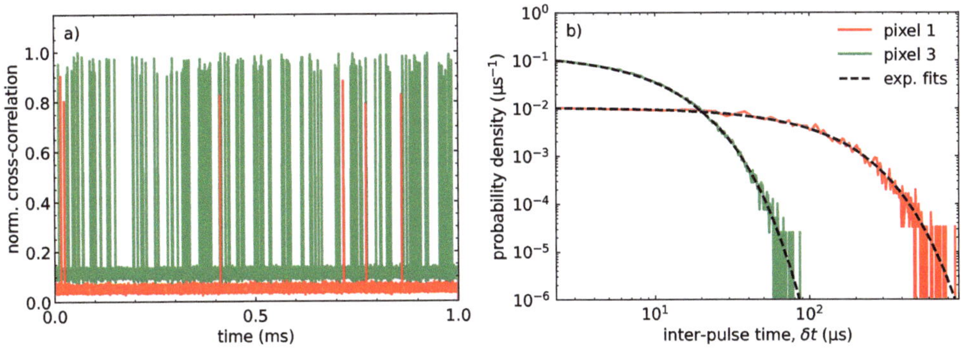

Figure 5.13: Photon-count statistics using a coherent laser diode as photon source and the matched-filter readout of two pixels of the RF-SNSPD array at different bias levels. (a) An exemplary, 1 ms-long window showing separate time traces of both pixels. (b) Probability density of inter-pulse time δt obtained from a 1 s-long time trace. The black, dashed lines are exponential fits by Eq. 5.6 to the corresponding curve.

the respective measured pulse templates. The frequency of the simulated pulse was then swept between these two limits, and a matched-filter output peak of half the value of the adjacent pixel (in this case pixel 3) was taken as the maximum value in order to still being able to distinguish between both pulses. With this criterion, the minimum f_r separation was simulated to be $\approx$30 MHz. Assuming an unchanged loaded bandwidth of 20 MHz per pixel and smaller separation bands of 30 MHz between them, a total number of 100 pixels per channel could be realized within a bandwidth of 5 GHz and read out by matched filters.

Beyond attributing single detection events to the correct pixel, we derived photon-count statistics with this method for the two simultaneously biased pixels over a longer acquisition time. To this end, a 1 s-long time trace of detector events on the transmission line was recorded. Both pixels were flood-illuminated by the 850 nm VCSEL diode. In order to compare count statistics at different η_{int}, both pixels were intentionally biased at unequal bias points relative to I_{sw}, with pixel 1 operated at much lower RF bias power than pixel 3. The recorded trace was then correlated with the individual pulse templates of both pixels, which results in two separate time traces (see Fig. 5.13a). The pulse-arrival times of each trace was obtained by setting the trigger level to half the maximum pulse height. From these traces, we derived the probability density of time δt between two sequential pulses, which is shown in Fig. 5.13b.

The coherent photon emission from the laser diode resembles a Poisson process and the probability density function of the photon inter-arrival time is expressed by

$$f(\delta t) = \gamma e^{-\gamma \delta t} \tag{5.6}$$

with mean photon rate γ. For accurate measurements, it is important that the photon rate is low enough such that the detector's dead time is significantly smaller than the photon inter-arrival time. As seen from fits to both curves with fitting parameter γ in Fig. 5.13b, the inter-pulse distributions of both pixels follow the expected photon distribution very well, which means that both detectors operate in single-photon regime.[4] The mean detection rates amount to $\gamma_{\mathrm{px1}} \approx 10\,\mathrm{ms}^{-1}$ and $\gamma_{\mathrm{px3}} \approx 138\,\mathrm{ms}^{-1}$, respectively. The large difference of γ between pixel 1 and 3 stems from the different bias levels and therefore η_{int} between these two pixels. Nevertheless, both probability densities of δt in Fig. 5.13b are well described by Eq. 5.6, independent of the specific bias point.

5.4.3 Future readout of RF-SNSPD arrays

For the proof-of-principle demonstration, the results herein were obtained by two separate analog signal generators as bias sources and post-processing on recorded time traces. In the future, the operation and readout of RF-SNSPD arrays could be implemented in real time on a software-defined radio (SDR) platform, similar to what has been demonstrated for MKID arrays [253].

The probe tones matching the resonance frequencies of an extended RF-SNSPD array would be generated by an FPGA-based AWG. After digital-to-analog conversion, the tone comb can be up-converted into the appropriate frequency band. The amplified output signal of the array would then be down-converted into the IF band and digitized. To extract the spatial information of each pulse from its IF-band frequency, either an FFT-based detection scheme or matched filters running on FPGAs could be used.

It remains to be seen how the extraction of timing information from the matched-filter output affects the timing jitter. In Ref. [254], a matched-filter approach was adopted to improve the readout of an SNSPD array with varying pulse height. With pulse-arrival times taken at the peak of the matched filter outputs (at the highest SNR), the timing jitter could be slightly improved compared to a conventional readout. For RF-SNSPD, however, it could also be necessary to extract timing information directly from the RF signal and to only determine the corresponding spatial information from the matched-filter output.

[4] Similar photon statistics recorded with SNSPDs have been presented in Refs. [49, 73, 252].

5.5 Summary

In this chapter, the concept of RF-SNSPD arrays, previously introduced by Doerner *et al.* [186], was investigated further, mainly with regard to the consequences of the oscillating bias current for the detector operation.

The RF bias leads to a stable operation without the risk of the detector latching into the resistive state. The self-resetting behavior is the consequence of the limited hotspot growth because of the alternating current through the nanowire and the current redirection out of the nanowire due to the damped resonance in the case of a finite hotspot resistance. Measurements of the bias-dependent dark-count rate show that even if the detector is biased by a current with an amplitude above I_{sw}, the device does not latch but exhibits self-oscillations at a pulse rate proportional to the current amplitude. In addition, an RF-biased nanowire with a ringdown time of only 0.4 ns was simulated and compared to DC bias at high photon-count rates using an electrothermal model of Kerman/Berggren [53, 60]. While the DC-biased operation led to relaxation oscillations and eventually latching due to back-biasing caused by the capacitive readout, the RF bias enabled a reliable restoration of the superconducting state after every detection event.

Another consequence of the RF bias is the gated detector operation because the microwave current modulates the instantaneous intrinsic detection efficiency $\eta_{int}[I_{rf}(t)]$. This was confirmed by analyzing the pulse envelopes when triggering on both rising edges of the RF pulses, showing a gated detection at the frequency $f_g = 2f_r = 3.8\,\text{GHz}$. Although gating is not necessary for the SNSPD operation, in applications with known photon arrival time spans, a gated operation is beneficial since background noise can be suppressed and the signal-to-noise ratio therefore improved. Synchronizing the RF-SNSPD to a fast VCSEL diode modulated with $f_{opt} = f_g$ revealed phase-sensitive changes of the photon-count rate with a modulation index of 0.71, which agrees well with a simple model of the operation conditions. In this way, we showed that the detection efficiency of RF-SNSPDs can be maximized and the influence of stray light minimized (here by a factor $\approx$6.0) by using a phase-locked light modulated with the detector's gating frequency. However, the oscillating intrinsic detection efficiency also implies a reduced duty cycle of the detector for photons with random arrival times, depending on the bias-dependent count rate. In our particular case, the average detection efficiency amounted to 30 %. For devices with a pronounced count-rate plateau, the duty cycle of the detector can be much larger and tuned over a broad range by setting the bias amplitude.

Lastly, a proof-of-concept readout of an RF-SNSPD array using pre-recorded templates of each pixel's response as matched filters showed the reliable attribution of each RF pulse on the common transmission line to the respective pixel, thereby enabling spatial resolution of the count events. Photon-count statistics over 1 s, extracted from two pixels simultaneously, revealed an exponential distribution of the inter-pulse time in accordance with the Poissonian photon-number distribution of the coherent light source.

6 Optical coupling

By its nature, a nanowire only covers a small area, which makes efficient photon guiding a challenge. Typically, SNSPDs consist of meandered nanowires covering a larger area on the order of several µm side length, which, however, requires long nanowires. This impairs the main beneficial properties of SNSPDs in timing and maximum count rate. Long nanowires are also more prone to constrictions, which potentially limit the spectral bandwidth of the detector. Therefore, the nanowire should be as short as possible while still allowing for efficient optical coupling from the system input. This trade-off not only affects free-space but also fiber coupling. To account for typical mode-field diameters of a telecom single-mode fiber ($\approx$10 µm) and a potential misalignment between the fiber core and the detector, the nanowire is typically >1 mm long [252].

In Section 2.6, existing methods to improve the coupling efficiency for SNSPDs have been presented, for free-space as well as for fiber and on-chip coupling. In this chapter, the integration of SNSPDs with free-form polymer structures will be investigated in order to achieve high coupling efficiencies with comparatively short nanowires.

These polymer structures are fabricated by 3D direct laser writing (DLW), a technique which uses two-photon polymerization of a photoresist in the focal region of a near-infrared, femtosecond-pulsed laser. By moving the focal region of the beam through the resist, the material gets selectively polymerized within a small volume (voxel) of typically <1 µm in diameter. Because of the squared intensity dependence of the photon absorption rate in the resist and therefore also the exposure dose, arbitrary 3D structures can be written with high accuracy even in axial direction (see the explanation in Ref. [255]). An overview of this technology is given in Refs. [256, 257]. Using 3D DLW, beam-shaping elements [193, 258] and optical interconnects [201, 202, 204, 205] have been demonstrated before.

Typically, these structures are used for applications at room temperature, e.g., in photonic circuits. Extending the 3D-DLW technique to optical coupling of SNSPDs (or superconducting photon detectors in general) poses questions about the mechanical stability of the polymer structures during cryo-cycling, potential harm to the detectors during laser lithography, and, ultimately, the achievable coupling efficiency.

The investigation of SNSPDs integrated with these optical elements is separated in two parts of this chapter. Section 6.1 deals with microlenses on top of an RF-SNSPD array, mainly targeting free-space coupling. As outlined in Section 2.5.2, RF-SNSPDs have the drawback of a low fill factor of the photon-sensitive area and could therefore particularly benefit from enlarged light-collection areas. In Section 6.2, photonic wire bonds will be explored for the coupling of a single-mode fiber array to a DC-biased SNSPD array consisting of detectors with comparatively short (≈105 µm long) nanowires.

6.1 Microlens-integrated RF-SNSPD

The research on microlens-integrated SNSPDs was conducted in collaboration with the Institute of Photonics and Quantum Electronics (IPQ) at KIT. An analysis of the lens design and experimental data on a DC-biased SNSPD were published in:
[259] Y. Xu, A. Kuzmin, E. Knehr, et al., "Superconducting nanowire single-photon detector with 3D-printed free-form microlenses", Optics Express ***29****, 27708 (2021).*
In addition to the results presented there, the same lens concept was also applied to RF-SNSPDs and is shown in this section. Y. Xu designed and fabricated the lenses. A. Kuzmin and E. Knehr performed the lens characterization and analyzed the experimental data.

In free-space applications of SNSPDs, photons are usually impinging the detector with an incidence normal to the substrate plane. The focussing optics and alignment method of the detection system therefore directly determine the area the nanowire has to cover for an efficient coupling.

In Ref. [259], we demonstrated the feasibility of integrating SNSPDs with polymer microlenses fabricated by 3D DLW. These lenses focus incident photons onto the active area of the detector in the focal plane of the lens. The effective collection area is increased from the detector area A_d to the collection area of the lens $A_\mathrm{c} = \pi r_\mathrm{c}^2$ with radius r_c. The optical coupling efficiency is not bound to the length of the nanowire anymore and the microlenses therefore enable the combination of a short nanowire with a large light-collecting area. The effective lens gain $G_\mathrm{d} = A_\mathrm{c}/A_\mathrm{d}$ could be arbitrarily increased by scaling the lens dimensions. However, a larger physical size of the lens goes along with a larger lateral shift of the focal point for light with an incidence direction nonparallel to the optical axis of the lens (see Eq. 9 in Ref. [259]). Depending on the detector size and the tolerable angular misalignment, this might set an

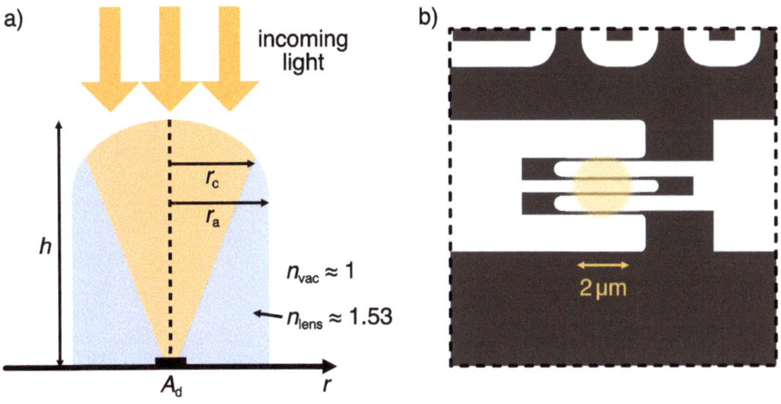

Figure 6.1: Microlens concept and dimensions. (a) Side-view sketch of a microlens (rotationally symmetric around the optical axis denoted as a dashed line) which focuses incoming light with incidence normal to the detector plane onto the active area A_d. The lens has a total height of $h = 70\,\mu m$ (equivalent to focal length f), a physical radius $r_a = 30\,\mu m$, and an effective collection radius $r_c \lesssim r_a$. (b) Top view of the active area A_d of an RF-SNSPD pixel and approximate dimension of the spot size in the detector plane.

upper limit to the feasible lens gain. In a proof-of-pinciple device, we demonstrated a gain of approximately two orders of magnitude for a lens diameter of 60 µm and an SNSPD area of 4.5 µm × 4.5 µm, which is reasonably consistent with the simulated value of $G_d = 117$. The experiment was conducted using two DC-biased SNSPD in close distance, one with a lens on top and the other as a reference detector without a lens.

In this section, the lens concept is applied to frequency-multiplexed SNSPDs. Beyond the importance of short nanowires for DC-biased SNSPDs, increasing the effective light-collection area for these RF-SNSPD arrays is of particular importance because the active areas are small compared to the size of each pixel, which includes the large resonator. As described in Section 2.5.2, the length of the nanowire is limited due to the requirement $L_k \ll L_r$ in order to have a resonance frequency which is near-independent of the bias-dependent kinetic inductance $L_k(I_b)$.

In the following, the basic lens concept and fabrication are briefly described before going into detail of the experimental characterization of a microlens-integrated RF-SNSPD.

6.1.1 Microlens concept and fabrication

A sketch of the lens concept is shown in Fig. 6.1a. The polymer has a refractive index of $n_{lens} \approx 1.53$ at 1550 nm (VanCoreA, Vanguard Automation GmbH). Incoming light within the

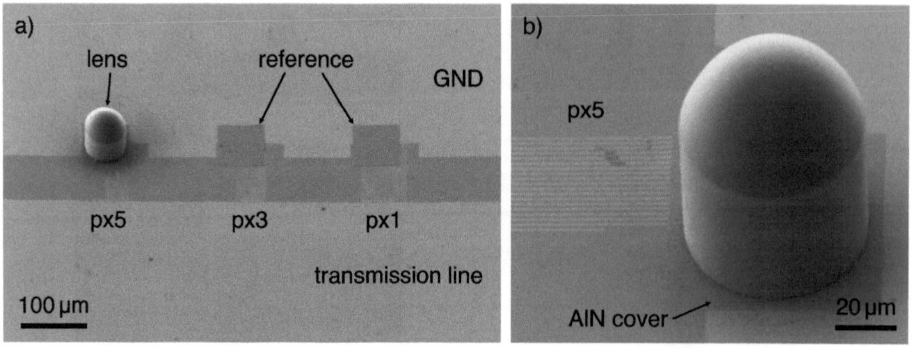

Figure 6.2: SEM micrograph of the RF-SNSPD sample. (a) SEM micrograph of the RF-SNSPD array with the examined pixels 1, 3, and 5. The first two serve as reference detectors while the latter has a lens on top. Each active area and part of the resonant circuit is covered by a square AlN cover. Light gray areas show the NbN layer. (b) Close-up of the microlens on pixel 5.

radius r_c is refracted onto the detector of size A_d. In general, $r_c \lesssim r_a$ (r_a: physical radius of the cylinder) because of Fresnel reflections at steeper angles of the incident light with respect to the lens surface [259]. A detailed analysis of lens shapes is given in Section 3 of Ref. [259]. The particular design used here (identical to the one in the experimental section of Ref. [259]), is a polynomial approximation of a spheroid with a minor half axis length of 32 µm, clipped at $r_a = 30\,\mu m$. The lens height h is designed to match its focal length $f = h = 70\,\mu m$.

In Fig. 6.1b, the approximate focal-spot size (2 µm diameter) which follows from wave-propagation simulations (see Fig. 3b in Ref. [259]) is visualized on the nanowire of the RF-SNSPD pixel. The simulated lens gain in Ref. [259] of $G_d = 117$, obtained for $A_d = 4.5\,\mu m \times 4.5\,\mu m$, corresponds to an effective collection radius of $\approx 27.5\,\mu m$ and $A_c = 2376\,\mu m^2$. Here, the nanowire covers a smaller area of $4.0\,\mu m \times 2.1\,\mu m$ which leads to a higher expected gain of 283.

The calculated value of G_d is experimentally proved by comparing photon-count rates of a lensed and a non-lensed detector. Instead of two separate, DC-biased SNSPDs, however, we use multiple detectors of a five-pixel RF-SNSPD array. The array is the same one as used before in Section 5.4. For the comparison, we used pixel 5 with a lens and pixel 1 and 3 as reference detectors (see also $|S_{21}|\,(f)$ plot in Fig. 5.10).

The basic fabrication process of the RF-SNSPD array is described in Section 5.1. Additionally, the active area of each pixel is covered by a 20 nm thick aluminum nitride (AlN) layer, which is fabricated by room-temperature sputter deposition of aluminum in a nitrogen atmosphere and lift-off. These covers serve as passivation of the nanowires and protection in the subsequent lens-fabrication steps. AlN is electrically insulating and is transparent for photons from UV to mid-IR [260]. The lithography of the lenses is described in Section 4.1 of Ref. [259]. In short: the photoresist is applied by drop-casting onto the detector chip, the precise position of the nanowire is acquired by optical recognition of alignment marks close to the detector, and the structure is written by moving the laser spot via galvanometer mirrors and a piezo stage through the resist. As an important measure to prevent degradation of the nanowire, we left a small cylindrical volume of resist of height 3 µm directly on top of the nanowire unexposed. Nanowires directly exposed by the focused laser were found to degrade, as seen by a $\approx$5 K lower T_c and much higher room-temperature resistance. After development in propylene glycol methyl ether acetate (PGMEA) and flushing in isopropanol (IPA), the unexposed volume was solidified by post-exposure using a UV lamp. SEM micrographs of the three pixels characterized in the following section are shown in Fig. 6.2. The active areas with a pitch of 200 µm are below square AlN covers with a side length of 70 µm.

6.1.2 Optical measurements of the microlens gain

To characterize the sample, the setup for bias and readout of the array is in accordance with the setup in Fig. 5.5. The array is broadly illuminated with normal incidence by a single-mode fiber (SMF-28) fixed at a distance of 40 mm from the detector plane (Fig. 6.3a). A near-plane-wave illumination onto the sample is assumed in this configuration. Because of the comparatively small distance of 200 µm between pixel 3 and 5 (400 µm between pixel 1 and 5), the photon flux incident on the lens and reference pixels is assumed to be the same. As the primary light source, we use a pulsed, femtosecond (fs) laser at 1550 nm wavelength and a repetition rate of 100 MHz.

To ensure identical intrinsic detection efficiencies of the pixels, the bias levels had to be calibrated. Two different methods were employed. For the first, here called DCR method, the bias was adjusted for equal dark-count rates of all three pixels. For the second calibration method (2λ-PCR), the array was illuminated from the back side through the sapphire substrate by a multimode fiber (Thorlabs FG105LCA) fixed 3 mm away from the chip. In this configuration, the sample can either be illuminated by the same fs-laser at 1550 nm or by a continuous-wave laser diode at 658 nm. A homogeneous illumination of all detectors is not guaranteed at this fiber-to-detector distance, which is, however, not necessary for the particular calibration method described later.

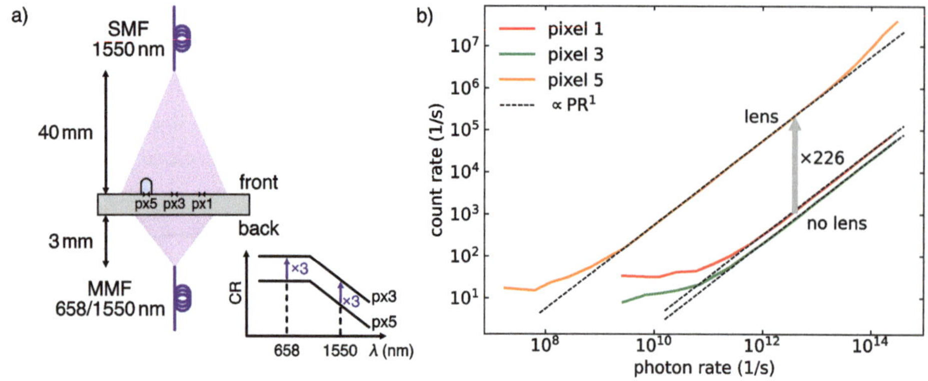

Figure 6.3: (a) Experimental configuration to test the lens performance. The detector chip can be illuminated from the lens/front side using a single-mode fiber at 1550 nm, or from the back side using a multi-mode fiber at 658/1550 nm. Using the latter illumination, the bias power of each pixel can be adjusted to obtain parallel CR(λ) curves, as visualized. (b) Comparison of count rate vs. photon rate curves between detectors without a lens (pixels 1 and 3) and a detector with a lens on top (pixel 5). The dashed black lines are fits to $\propto$ PR^1. The measured count rate increase for the lens sample is $\approx$226 compared to the average count rate of pixel 1 and 3 at the same incident photon rate.

For the DCR-calibration method, the sample was only illuminated from the front/lens side. The lens gain can be obtained by comparing photon-count rates of the detector with a lens to the reference detectors without a lens:

$$G_\mathrm{d} = \frac{\eta_\mathrm{oc,lens}}{\eta_\mathrm{oc,ref}} = \frac{\mathrm{PCR}_\mathrm{lens}}{\mathrm{PCR}_\mathrm{ref}}. \tag{6.1}$$

The latter equality is true if the incident photon rate PR (see above), the absorption η_abs, and the intrinsic detection efficiency η_int of the lens and reference detectors are equal (see Eq. 2.9). The absorption is assumed to be identical because of the parallel fabrication of the detectors from the same film. Efficiency η_int at a given λ depends on the relative bias of the detectors. At 1550 nm wavelength, the bias-dependent count rates did not saturate, thus $\eta_\mathrm{int} \ll 1$. To approximate equal η_int across all pixels under this condition, the bias power of each detector was adjusted in such a way that the dark-count rates are roughly equal. Because all pixels have nominally identical nanowire geometries with the same number of turns, it is assumed that the intrinsic detection efficiency in this case is also comparable.

Count rates over incident photon rates were recorded for each pixel. To this end, each pixel was biased individually by a single-tone bias signal and measured one after another. These curves are plotted in Fig. 6.3b. From proportionality CR $\propto$ PR (fits as dashed, black lines), it is seen that all curves exhibit a broad single-photon regime. Toward low incident photon rates, the count rates deviate from this proportionality and saturate at the DCR level. At high photon rates, the curve of the lens sample shows a higher-order dependence on PR, indicating an incipient multi-photon regime.

From this graph, the lens gain can be estimated by comparing photon-count rates at the same photon rate in single-photon regime (Eq. 6.1), leading to $G_\mathrm{d} \approx 226 \pm 53$. The difference between reference pixel 1 and 3 might be the result of different η_int as the DCR level was not constant: $\mathrm{DCR}_\mathrm{px1} \approx 30\,\mathrm{Hz}$, $\mathrm{DCR}_\mathrm{px3} \approx 10\,\mathrm{Hz}$, and $\mathrm{DCR}_\mathrm{px5} \approx 20\,\mathrm{Hz}$.

For the 2λ-PCR method, the same sample was illuminated from the back side through the sapphire substrate to calibrate η_int of pixel 3 and 5 more accurately. Using back-side photon incidence at two wavelengths, 658 nm and 1550 nm, the bias power levels of pixel 3 (reference) and 5 (lens) were adjusted to achieve the same PCR ratio at both wavelengths:

$$\frac{\mathrm{PCR}_\mathrm{lens}(1550\,\mathrm{nm})}{\mathrm{PCR}_\mathrm{ref}(1550\,\mathrm{nm})} = \frac{\mathrm{PCR}_\mathrm{lens}(658\,\mathrm{nm})}{\mathrm{PCR}_\mathrm{ref}(658\,\mathrm{nm})}. \tag{6.2}$$

In this case, it is expected that both pixels are operated at the same intrinsic detection efficiency if η_abs and the ratio of incident photon rates on both detectors, $\mathrm{PR}_\mathrm{lens}/\mathrm{PR}_\mathrm{ref}(1550\,\mathrm{nm}) = \mathrm{PR}_\mathrm{lens}/\mathrm{PR}_\mathrm{ref}(658\,\mathrm{nm})$, are both independent of the wavelength (see also Eqs. 30–32 in Ref. [259]). This is illustrated by the parallel $\mathrm{CR}(\lambda)$ curves in the inset of Fig. 6.3a. Both the cut-off wavelength λ_c as well as the slope of the curves at $\lambda > \lambda_\mathrm{c}$ depends on the bias (see also Fig. 4.4 in Section 4.3.2). At the same relative bias, both count-rate curves are parallel and the vertical shift is only caused by different (absolute) incident photon rates on both detectors.

After calibration of the bias levels using back-side illumination, the optical setup was switched back to front-side illumination to compare count rates. At $\mathrm{PR} = 1.83 \times 10^{12}\,\mathrm{Hz}$, count rates of 2765 Hz (reference) and 678 134 Hz (lens) were measured. This corresponds to a lens gain of $G_\mathrm{d} = 245$, which fits well to the mean value of 226 obtained from the previous measurement. The experimentally obtained values of the lens gain are also in quite good agreement with the previously mentioned theoretical value of $G_\mathrm{d} = A_\mathrm{c}/A_\mathrm{d} \approx 283$, thus matching the increase in effective collection area by the lens.

Weak points of the particular method used here to experimentally obtain the lens gain are the assumptions of a constant intrinsic detection efficiency η_int and constant front-side incident photon rate PR between the lens and reference detector. While the former could be ensured more stringently by detectors with a wide saturation in $\mathrm{CR}(I_\mathrm{b})$ (e.g., by operating at lower temperatures), the latter could be substantiated by measurements of the beam profile in the detector plane. Nevertheless, the experimental lens gain is in the expected range and fits well to the measurements in Ref. [259].

The results of this section show the feasibility of microlens-integrated RF-SNSPD arrays and thereby a method to overcome the conflict of achieving a high coupling efficiency with very short nanowires. The proof-of-principle device discloses a clear path to substantially increase the low fill factor of previously shown RF-SNSPDs arrays, mainly achievable by reducing the pixel pitch and increasing the lens diameter.

The pitch of the array and the lenses could also be adjusted to match common fiber arrays, e.g., 127 µm. This device could be combined with microlenses fabricated on the facets of a fiber array (see, e.g., Ref. [193]) to facilitate fiber-to-detector coupling. The main advantage would be the larger beam waist which relaxes the requirements regarding the fiber-to-detector alignment. However, this would still require an active alignment of the components, for example by maximizing the detector counts. A method to directly connect fiber cores to SNSPDs will be explored in the following section.

6.2 Optical packaging of SNSPDs with photonic wire bonds

As with the lens-integration, this section is the result of a collaboration with IPQ. Y. Xu, P. Maier, and Y. Bao designed and fabricated the PWBs. M. Ziegler (IPHT) deposited the NbN films. E. Knehr designed and fabricated the SNSPD chips and performed the experiments. Initial tests of PWB-SNSPDs (not included here) were part of a master's thesis:
[261] C. E. Jimenez, "Fiber-coupled superconducting nanowire single-photon detector using photonic wirebonds", master's thesis (Karlsruhe Institute of Technology, 2020).

Photonic wire bonds (PWBs) are optical interconnects in the form of 3D polymer waveguides which guide photons in the visible and near-infrared range via the refractive index contrast to the surrounding, low-index cladding material (here: air/vacuum). PWBs have previously been demonstrated as a flexible method for source-to-chip [204] and chip-to-chip coupling [201] for silicon photonic circuits. Here, the focus is on the peculiarities that come with the integration of PWBs with cryogenic detectors: the fabrication on top of the detector while avoiding degradation of the nanowire, the mechanical stability of the bonds at low temperatures, and the achievable coupling efficiency. Multiple package arrangements between the fiber and the detector chip are possible, in principle. For a proof-of-principle demonstration of the concept, we chose to place the fiber array at the edge of and in parallel to the detector chip. In the following, the detector layout and fabrication is presented before briefly outlining the

PWB design and lithography. In Section 6.2.3, a proof-of-principle PWB-SNSPD device will be characterized under illumination and compared to a reference detector.

6.2.1 Detector design and fabrication

To demonstrate the PWB-coupling method for SNSPDs, a chip layout was designed to match a commercial, lidless fiber array consisting of eight fibers in a row with a pitch of 127 µm. The fibers have a mode-field diameter (MFD) of ≈10.4 µm and are single-moded at 1550 nm wavelength.

The detector design has to account for the limitations given by the lithography system used to write the bonds. The piezo stage and galvanometer scanner limit the structure size to 300 µm × 300 µm × 300 µm. Therefore, both interfaces (fiber core and detector) have to be placed within this volume. In the chosen arrangement with the fiber array in parallel to and close to one edge of the detector chip, the detectors need to be placed close to this chip edge. To account for additional clearance between the chip edge and fiber array, the detectors were placed at a distance of 110 µm from the edge. Directly writing fine structures by electron-beam lithography (EBL) so close to the chip edge is difficult due to the uneven resist thickness close to the chip border after spin coating. Therefore, several detector chips (3.3 mm × 3.3 mm) were patterned on a bigger wafer (10 mm × 10 mm) and precisely diced afterwards.

The size of the active area of the SNSPDs is a compromise between the length of the nanowire and the mechanical stability of the PWBs. A short nanowire allows for higher count rates but covers a smaller area. The cross-section of the PWBs would therefore have to be smaller as well to match the detector. We chose an active area of 4.5 µm × 4.5 µm with a fill factor of 0.4 (80 nm wire width, 200 nm pitch). This way, the bonds can be several µm wide (in the following section designed to have a cross-section of 4.0 µm × 4.0 µm) while the nanowire is still relatively short at $l \approx 105\,\mu m$.[1]

In order to enable an automated acquisition of the detector positions, crosses around the active area serve as alignment marks which can be detected by the two-photon lithography system. For bias and readout, the detectors were connected to contact pads at the chip edges by coplanar waveguides (CPWs) with a characteristic impedance of 50 Ω. The layout of the SNSPD chip is shown in Fig. 6.4.

[1] It should be noted that, with a side length of several µm, the bonds become multi-moded at the target wavelength of 1550 nm. In the intended configuration, however, this should not affect the coupling efficiency.

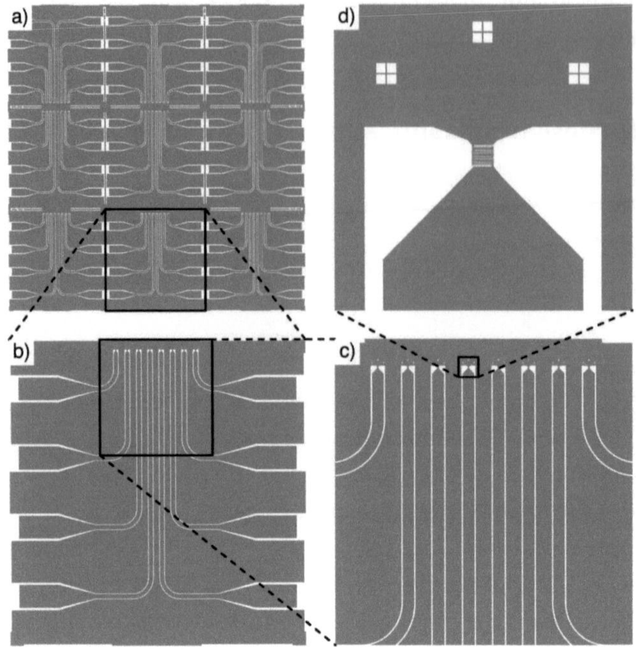

Figure 6.4: Design of the SNSPD chip for PWB coupling. (a) 10 mm × 10 mm wafer with nine identical detector chips. (b) 3.3 mm × 3.3 mm detector chip with eight detectors along the upper edge, which are connected to contact pads (left and right side) by CPWs. (c), (d) Details of the detectors. Each active area has a size of 4.5 µm × 4.5 µm. To align the PWBs, the exact position of each nanowire is acquired from the three alignment marks visible above the detector in (d).

To fabricate the detectors, 5 nm thick NbN was deposited by ALD on 10 mm × 10 mm quartz substrates. Quartz was chosen because its refractive index of ≈1.46 is smaller than that of the PWB polymer at 1.53. While this is less important for the chosen configuration with normal photon incidence on the detectors, the use of quartz substrates would potentially also allow for evanescent coupling in the future. The latter would be prevented when using, e.g., silicon ($n \approx 3.45$) or sapphire ($n \approx 1.77$) substrates [262]. The properties of ALD-NbN films on quartz have been investigated in Section 3.2. The detectors are defined by EBL in two steps, in accordance with the patterning process described in Section 4.1.2. The nanowires and alignment marks are defined in a negative-tone process, while the rest of the layout (CPWs, contact pads, and dicing marks) is exposed in a positive-tone process. After lithography and development, the chip is patterned in a single etching step using ion milling. Lastly, the remaining resist is removed by submersion in acetone and the 10 mm × 10 mm substrate is diced into nine chips with eight detectors each.

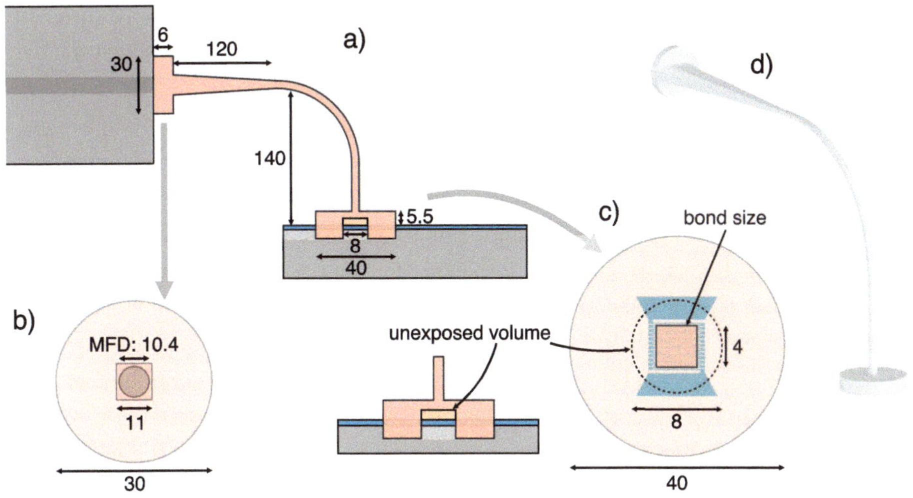

Figure 6.5: Sketch of the PWB geometry and dimensions (unit: µm; sketch dimensions not to scale). (a) Side view of the bond from the fiber core (upper left) to the detector chip (lower right). (b) Front view of the fiber interface. The taper starts with a square cross-section of 11 µm × 11 µm and is fixed to the fiber (10.4 µm MFD) by a cylindric anchor of 30 µm in diameter. (c) Top view of the detector interface. The bond size of 4 µm × 4 µm is sketched as a square on the slightly larger active area of the SNSPD underneath. The cylindrical volume left unexposed during the lithography step is marked as a dashed circle (8 µm in diameter) and in the side view (3 µm in height). (d) 3D render of a photonic wire bond.

In previous experiments with microlenses (Section 6.1), we covered the active detector areas by a 20 nm thick AlN layer to protect the nanowire in the subsequent fabrication steps. Later it turned out, however, that the unexposed volume in direct vicinity of the nanowire is enough to prevent degradation. Here, we omitted the additional fabrication of AlN covers.

6.2.2 Direct laser-writing of the bonds

The PWB design is sketched in Fig. 6.5. The structure consists of cylindrical anchors on both interfaces to ensure a proper fixation to each interface, a taper section to adapt the MFD, and the actual bond with a cross-section of 4 µm × 4 µm and a 90° bend. At the fiber side, the taper has a square cross-section of 11 µm × 11 µm to encompass the whole mode field from the fiber. The cross-section is then tapered down over a straight, 120 µm long section to the size of the bond on the detector side. The bend has a rather large radius of ≈80 µm to minimize bending losses.

For fabrication, the detector chip and the fiber array were glued by thermally cured silver paste onto a common aluminum carrier. The carrier is designed to roughly adjust the height of the detector substrate and the quartz in which the fibers are embedded in. The vertical distance between the detector surface and the fiber core of ≈140 µm leaves enough space for the 90° bend. When placing both components on the carrier, an alignment precision of few 10s of µm is sufficient and easily achievable by common flexure stages. Thereafter, the photoresist was applied by dropcasting onto the sample. As a resist, we used the same negative-tone resist ($n = 1.53$) as before in Section 6.1. Next, the exact position and direction of each interface was acquired. The fiber core can be detected by utilizing fluorescent emission by the photoresist at a much lower laser power than what is used for two-photon polymerization. This emission can be detected through the fiber and maximized by scanning the focal region through the resist in the vicinity of the fiber interface. The fiber-core detection scheme is described in Ref. [263]. The SNSPD position is obtained from the alignment crosses close to the nanowire. With both positions acquired, the exact trajectory of the bond is calculated. Along the trajectory, the contours of the different bond parts are laid out to get a solid 3D structure. This structure can be exposed by different writing strategies. Here, we use a bottom-to-top method for which the structure is sliced into individual layers along the z-axis and each layer is written by hatching in the xy-plane. The voxel is meandered with a hatching distance of 100 nm before moving up in z-direction in steps of 100 nm. With voxel dimensions of 0.6 µm × 0.6 µm × 1.6 µm (xyz), there is a large overlap between the individual lines and layers. The larger diameter in z-direction is compensated for by shrinking the layout in the same direction in order to obtain a structure with dimensions as originally designed. With the particular setup and lithography parameters, the writing time per bond is around 20 min.

Similar to the lens fabrication in Section 6.1.1, a small cylinder with a 8 µm diameter and a height of 3 µm in the direct vicinity of the nanowire is left unexposed during the lithography to prevent degradation of the superconducting structure. After exposure, the resist is developed by submerging the sample in PGMEA and flushed using IPA. A post-exposure step following afterwards is applied using UV light at 500 mW/cm^2 for 40 s to solidify the remaining liquid volume.

An SEM micrograph of the sample is shown in Fig. 6.6. All eight bonds are intact and without any visible damages. An additional anchor with a wrong shift from the alignment marks was erroneously written on the CPW of the rightmost detector. The remnants of the DLW resist on the detector chip (visible on top of the CPWs in the lower half of Fig. 6.6a) do not harm the detector operation and could be removed by thorough rinsing in IPA.

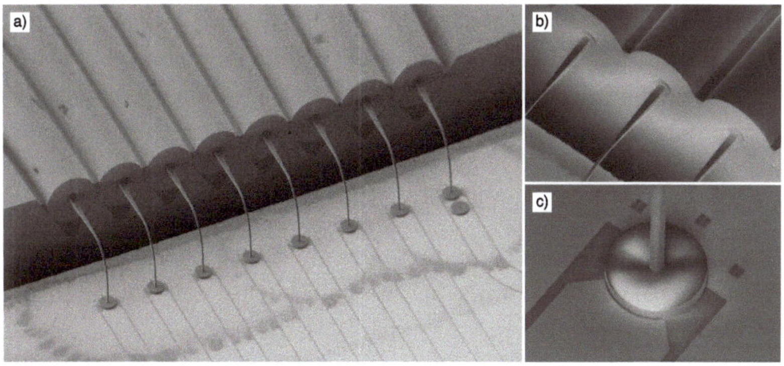

Figure 6.6: SEM micrographs of the PWB-SNSPD array. (a) Image including all eight PWBs. (b) Zoom on the PWBs attached to the fiber cores. (c) Zoom of the bond above the nanowire.

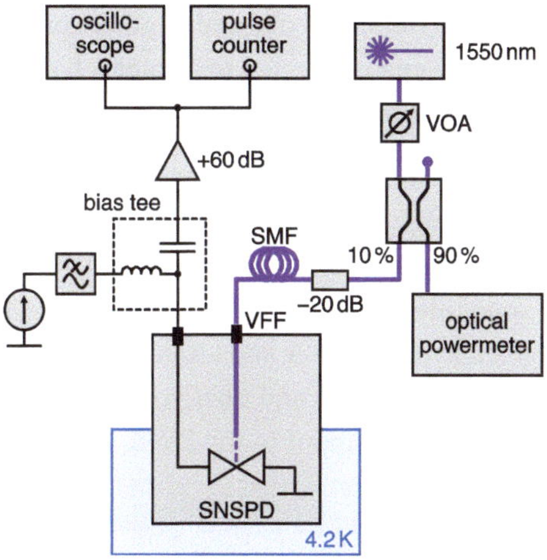

Figure 6.7: Experimental setup to characterize the PWB-SNSPDs. The optical path (violet) consists of a pulsed fs-laser at 1550 nm, a variable optical attenuator (VOA), a 90:10 fiber splitter, a fixed attenuator (−20 dB), a 2 m-long single-mode fiber (SMF), and a vacuum fiber feedthrough (VFF) to which the fiber of the PWB is connected.

6.2.3 Optical measurements of the PWB coupling efficiency

For measurements, the detector chip is connected on both sides to printed circuit boards (PCB) by aluminum wire bonds. The PCBs are then connected by SMP launchers to RF lines for bias and readout.

The measurement setup is depicted in Fig. 6.7. The PWB-SNSPD package is mounted in a vacuum dipstick which contains a small amount of helium (≈20 mbar) as contact gas and is submerged in LHe. The sample was cooled down to 4.2 K within approximately one hour. Over the course of the experiments, a total of four cool-down/warm-up cycles were performed. The PWBs stayed intact without any damage visible under the optical microscope after the experiments.

The SNSPD is biased by a current source through a low-pass filter and the DC port of a bias tee. The detector signal is read out by the RF port of the bias tee, amplified by about 60 dB and recorded by an oscilloscope and a pulse counter. With the cryogenic setup in use, only two detectors can be read out in a single cool-down. As a light source, we use a pulsed fs-laser at 1550 nm with the power tunable by a variable optical attenuator (VOA). To monitor the optical power, a 90:10 fiber splitter (Thorlabs TW1550R2A1) with a bandwidth of 200 nm around 1550 nm is used. The optical power is measured at the 90 % port using a photodiode (Thorlabs S154C) with a specified minimum detectable power of 100 pW. The 10 % output of the splitter is attenuated by 20 dB and connected to a single channel of the fiber array inside the dipstick via a vacuum fiber feedthrough (VFF).

Before the cryogenic measurements, the optical path was calibrated using a second powermeter connected to a very similar single-mode fiber (same MFD of 10.4 µm) which replaced the fiber inside the dipstick. The power ratio from the powermeter at the detector position to the monitor powermeter amounted to 1.081×10^{-3} (−29.66 dB). With an optimum ratio of 1.111×10^{-3} (−29.54 dB) given by the splitter and attenuator, we measure a loss of −0.12 dB which comes mainly from the feedthrough and fiber connectors. The measured power ratio was used to estimate the optical power input to the PWB.

Direct method

The main goal of the measurements, beside a demonstration of the general feasibility, is to obtain the coupling efficiency achievable with this coupling scheme. From Eq. 2.9, η_{oc} can be acquired by measuring the ratio of registered to incident photons. For the 5 nm thick NbN layer, $\eta_{\mathrm{abs}} \approx 0.3$ at 1550 nm wavelength [66], while η_{int} can be acquired from CR(I_{b}) curves.

Before the actual measurements under photon incidence, the detector was characterized with the laser turned off. Bias-dependent count rates of this characterization are plotted in the inset of Fig. 6.8a. With the input fiber disconnected from the dipstick and the vacuum fiber feedthrough optically blocked (see Fig. 6.7), a purely exponential DCR (black, solid line) is recorded as expected. After the optical setup was connected to the dipstick and with the laser still turned off, additional counts were registered (red, dashed line), leading to an additional

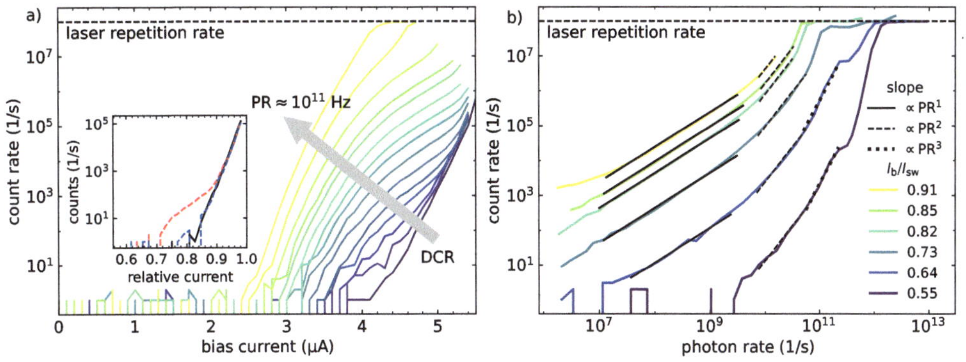

Figure 6.8: Count rate vs. bias current and photon rate of a PWB-coupled SNSPD. (a) Count rate over bias current for incident photon rates from 0 Hz (DCR) to $\approx 1 \times 10^{11}$ Hz. (Inset) Count rate over bias current with the fiber feedthrough closed (solid black), fiber connected and laser off (dashed red), and fiber connected and wrapped in aluminum foil (dashed blue). (b) Count rate over incident photon rate for various bias currents. Black lines are fits by the respective slope.

"hump" in the bias regime where the dark counts originating from intrinsic excitations are small. Only after wrapping the 2 m long fiber patch cable (connecting the fixed attenuator to the dipstick; labelled "SMF" in Fig. 6.7) in aluminum foil, the curve resembled an exponential bias dependence again (blue, dashed line). The source of the counts is therefore this fiber and not the rest of the room-temperature optical setup including the fiber splitter and attenuators. The most plausible explanation for the observed additional counts are background photons coupling into the standard fiber patch cable through the jacket (standard PVC jacket with 3 mm diameter). This influence was not observed in previous experiments (e.g., in Chapter 4) because of the inefficient coupling method via broad illumination used there. When coupled via PWBs, the presumably high coupling efficiency from the fiber to the active area together with a high intrinsic efficiency of the SNSPD at lower wavelength ($\leq$800 nm) seems to reveal this influence of stray photons.

In the main graph of Fig. 6.8a, the bias-dependent count rate is plotted for various incoming photon rates. The curve shifts from the purely exponential DCR (violet) close to $I_{sw} = 5.5\,\mu A$ to smaller bias currents at an increasing photon rate (yellow). The curves start with an exponential increase parallel to DCR (probabilistic regime, see Section 2.3.1) and begin to diverge from this exponential dependence at higher currents (in logarithmic scale). However, there is no saturation visible at bias currents close to the switching current, therefore $\eta_{int} \ll 1$. Moreover, the experimental data is not suited for a reliable fit by an error function according to Eq. 2.10. Hence, the measurement data at $\lambda = 1550$ nm and $T = 4.2$ K is not suited to estimate η_{int} and could therefore not be used to separate the individual efficiency factors in order to extract η_{oc} of the PWB. Hints of a high coupling efficiency are given by the stray light influence mentioned before and the qualitative dependence of switching current I_{sw} on the incident power. All curves in Fig. 6.8a were measured up to the switching current of the detector. Thus, one can see an influence on I_{sw} even at rather lower incident rates of PR $\approx$ 1 MHz (second curve from the right). This effect is studied quantitatively at the end of this section.

Measurements of the count rate over the incident photon rate are plotted in Fig. 6.8b. Depending on the number of incident photons, various detection regimes can be identified via the slope of the curves. At low photon rates $(1 \times 10^7$ to $2 \times 10^9)$ Hz, there is a broad single-photon regime with CR $\propto$ PR^1 (see fits as solid black lines in Fig. 6.8b). For higher photon rates, there are also regimes visible where two- and three-photon events dominate the count rate. At low bias currents, the absorption of a single photon is not enough to trigger a detection event and only multi-photon events are registered. The dents of the curve of the lower three bias currents can be explained by measurement artifacts due to changes in the pulse shape. At these high photon rates, changes in the pulse shape due to overlapping pulses were observed, resulting in reduced amplitudes. If the count rate is determined at a constant absolute threshold voltage, some pulses are not registered. When the incident PR is increased further, the pulse amplitude increases above the threshold again. All curves saturate when the laser repetition rate of 1×10^8 Hz is reached.

For bias currents $\leq 0.85 I_{sw}$, the DCR is almost negligible (see Fig. 6.8a), which is why only the yellow curve shows a plateau toward low photon rates. At $I_b = 0.91 I_{sw}$, the SDE amounts to $\approx 3 \times 10^{-4}$ in the single-photon regime. As explained for the bias dependence, the low SDE value is mostly limited by the intrinsic detection efficiency. All the measurements were conducted on the same detector (second from right in Fig. 6.6a) in the third cool-down. Another PWB-coupled detector was characterized in a subsequent cool-down, resulting in an almost identical SDE at the same ratio I_b / I_{sw}.

Reference method

Because the intrinsic detection efficiency could not be reliably estimated from the previous measurements, the optical coupling efficiency could not be determined individually from PCR/PR ratios. Therefore, the value of $\eta_{abs} \cdot \eta_{int}$ had to be acquired by separate calibration measurements. For this, we used the exact same detector which was previously characterized with PWB coupling but without the polymer structure on top and, instead, illumination with normal photon incidence from a fiber at a distance of 40 mm from the detector. In this configuration, the detector chip was broadly illuminated by a quasi-plane wave. The fraction

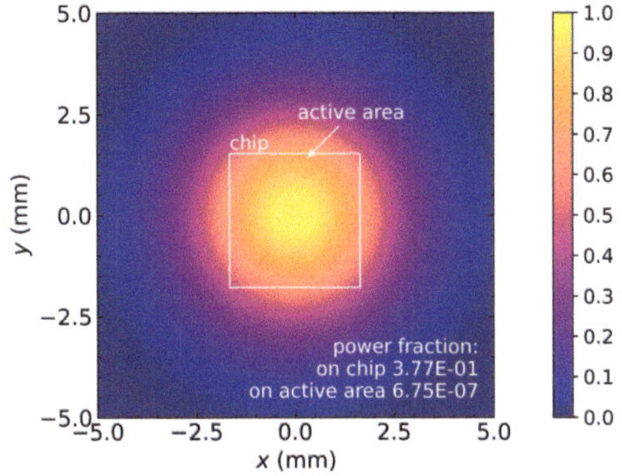

Figure 6.9: Simulated Gaussian beam intensity distribution in the detector plane at a distance of 40 mm from the fiber tip. The 3.3 mm × 3.3 mm detector chip is depicted as a white square. The position of the characterized detector with an active area of 4.5 µm × 4.5 µm is highlighted by the arrow.

of the optical power hitting the active area was estimated by Gaussian beam simulations. For this, we took a beam radius in the detector plane of (p. 43 of Ref. [264])

$$w(z) = \frac{\mathrm{MFD}}{2}\sqrt{1+\left(\frac{z}{z_\mathrm{R}}\right)^2} \tag{6.3}$$

with an MFD of 10.4 µm (SMF-28), a fiber-to-detector distance $z = 40\,\mathrm{mm}$, and Rayleigh length (Eq. 2.29 in Ref. [264])

$$z_\mathrm{R} = \frac{\pi}{\lambda}\left(\frac{\mathrm{MFD}}{2}\right)^2 . \tag{6.4}$$

With beam radius $w_\mathrm{f-d} = w(40\,\mathrm{mm})$, the detector-plane intensity distribution can be calculated as Gaussian distribution

$$I_\mathrm{beam}(x, y) = A \cdot \exp\left(-2 \cdot \left[\frac{x^2 + y^2}{w_\mathrm{f-d}^2}\right]\right) \tag{6.5}$$

in the xy-plane with amplitude A.

Figure 6.9 shows the calculated intensity distribution. The xy-position of the chip under the fiber tip (center of the beam) was measured with a precision of ≈20 µm and is highlighted in Fig. 6.9. In this range, the beam is distributed rather uniformly and uncertainties on the order of few 10 µm do not change the power fraction significantly. By taking the coordinates of the active area on the chip from the detector layout, the exact position of the detector under the beam is derived. The fraction of the total power which hits the active area of size

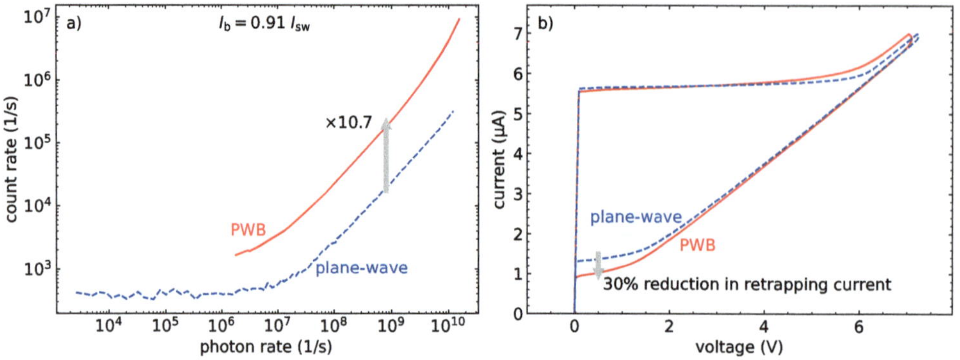

Figure 6.10: Comparison of PWB and plane-wave measurements of the same SNSPD. (a) Count rate over photon rate for PWB and plane-wave illumination at the same bias current. The PWB curve (red) is the same as the yellow curve in Fig. 6.8b. The plane-wave curve shows the experimental data which is shifted to lower photon rates according to the calculated power fraction hitting the active area. (b) Current-voltage curves of both cases without photon incidence.

4.5 µm × 4.5 µm was calculated to be 6.75×10^{-7} by integrating the simulated intensity over the respective area.

Using this coupling method with normal incidence, the detector was characterized in the same way as before. PWB and plane-wave measurements of the same detector are compared in Fig. 6.10. Graph (a) shows the CR(PR) curves in both cases for the highest measured bias current ($0.91\ I_{\mathrm{SW}}$). The blue, dashed curve is the result of the experimental data calibrated to the photon rate hitting the active area under normal photon incidence by taking into account the previously simulated power fraction (Fig. 6.9).

Comparing the count rates for PWB and plane-wave coupling at the same photon rate should directly point out the PWB coupling efficiency because η_{abs} is the same in both cases and, by the calibration using Gaussian beam simulations, $\eta_{\mathrm{oc,ref}} \equiv 1$. Because the curves were measured at the same relative (and absolute) bias current, the intrinsic efficiency η_{int} is assumed to be constant as well. Overlapping curves would signify a near-unity coupling efficiency by the PWB. Accordingly, a PWB curve below the plane-wave curve would illustrate the bond's coupling loss. However, as seen in Fig. 6.10a, count rates measured using the PWB actually exceed the ones of the reference in single-photon regime by a factor of ≈10.7.

One explanation of this result might be the thermal decoupling of the nanowire by the polymer structure on top of it. The sample is cooled by the surrounding contact gas, and the several µm thick polymer has a very low thermal conductivity at 4.2 K.[2] Increasing the phonon escape time would cause several observable changes in the detector characteristics: (i) a higher η_{int} (see, e.g., Fig. 14 in Ref. [38]), (ii) an increase in DCR, and (iii) a lower retrapping current. In Fig. 6.10a, an increase in DCR of ≈6 is indeed visible at low photon rates for which the PWB curve approximates a higher plateau. Comparing the current-voltage characteristic of both cases (Fig. 6.10b), a reduction of the retrapping current by ≈30 % is observed. This is expected in the case of a lower thermal conductance from the nanowire because I_r is the result of the balance between Joule heating and cooling of the normal domain. A reduced retrapping current therefore also reinforces this explanation. At the same time, I_{sw} is nearly identical. Slight differences in the shape of curve in the resistive state (Fig. 6.10b) are caused by a longer time constant of the low-pass filter which was used to record the PWB curve.

Another factor explaining the results of Fig. 6.10a could be the optical power which hits the detector chip in the normal-incidence measurement but does not contribute to the measured count rate. This optical power would heat the detector chip up. However, because the dark-count rate strongly depends on the temperature, it would have a growing influence on the measured count rate and would be seen by a super-linear dependence of CR(PR). Therefore, the observed dependence CR ∝ PR^1 at incident photon rates in the range (1×10^8 to 1×10^{10}) Hz implies a negligible photon-induced change in temperature of the chip. Uncertainties introduced by the simulations and the precise position of the detector under the fiber tip cannot be ruled out, but are unlikely to account for an error of an order of magnitude.

In any case, without a constant η_{int}, the observed factor of 10.7 would be the result of the optical coupling efficiency together with the increase in intrinsic detection efficiency:

$$\eta_{\text{oc,PWB}} \cdot \frac{\eta_{\text{int,PWB}}}{\eta_{\text{int,ref}}} = \left.\frac{\text{PCR}_{\text{PWB}}}{\text{PCR}_{\text{ref}}}\right|_{\text{PR=const.}} \approx 10.7. \tag{6.6}$$

Under the particular operation conditions, the reference measurement is therefore not suited to extract $\eta_{\text{oc,PWB}}$. One solution could be a reference measurement of a sample with only the cylinder printed on top of the detector to ensure a comparable thermal environment.

Switching-current method

Another method to estimate the coupling efficiency was employed by measuring the decreasing switching current when increasing the incident optical power. This effect is attributed

[2] We rule out a direct thermal load caused by the optical fiber from room temperature to 4.2 K because it is thermalized by the contact gas over the whole length in the dipstick.

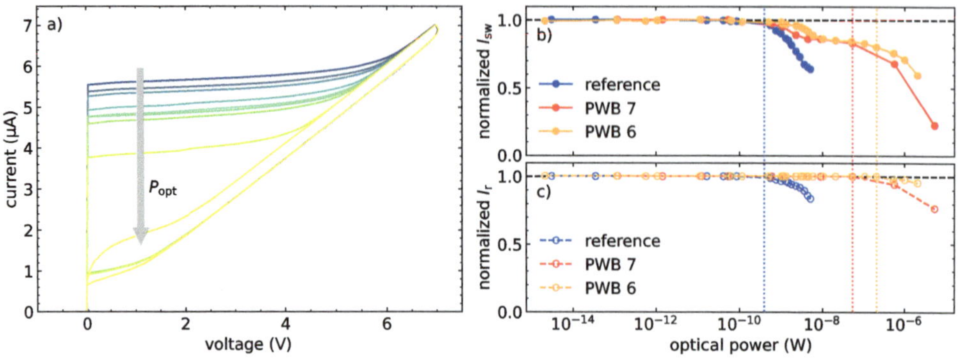

Figure 6.11: Current-voltage characteristics dependent on the incident optical power. (a) Current-voltage curves of a PWB-coupled SNSPD ("PWB 7") with increasing P_{opt}. (b), (c) Power dependence of switching (b) and retrapping current (c) of a reference detector (blue) and two PWB-coupled detectors (red, orange). All values are normalized to the maximum of the respective sample. The vertical, dotted lines denote the optical power levels at which the respective retrapping current starts to decline.

to a decrease in the effective superconducting cross-section when photons are absorbed. When the incident optical power is increased, the density of excited hotspots in the nanowire increases which leads to a decrease in effective cross-section and, thus, a decrease in the measured switching current.

Current-voltage curves depending on the optical power were recorded for two PWB-coupled samples ("PWB 6" and "7") as well as a separate reference sample ("reference") with nominally the same nanowire geometry. Sample "PWB 7" is the one characterized previously in this section. In the case of the reference sample, only the fraction of the incident optical power which actually hits the active area was taken into account, see the method from Eqs. 6.3–6.5 and Fig. 6.9. Comparing the samples without incident optical power, the retrapping current of the PWB samples at 0.95 μA and 1.18 μA is smaller than the one of the reference detector at 1.36 μA. This supports the previous assumption of an influence of the polymer on top of the nanowire on the thermal coupling to the environment, causing a lower retrapping current as well as a higher intrinsic detection efficiency.

In Fig. 6.11a, $I(V)$ curves of sample "PWB 7" are plotted for an increasing optical power (steps of P_{opt} are not equidistant). One can see a drastic decrease of I_{sw} with P_{opt}, while I_r is only reduced meaningfully at the highest measured optical power (yellow curve).

The power dependence of the relevant parameters, I_{sw} and I_r, is shown for all three samples in Figs. 6.11b and 6.11c. To compare only the relative dependence, all curves are normalized to the highest measured current of each of the three samples. Comparing I_{sw}, one can see a power-independent value up to 3×10^{-10} W. At higher power levels, the switching current of the reference detector decreases linearly (plot in linear-log scale), while I_{sw} of the PWB samples first decreases with a shallower slope before almost flattening. Only at even higher power levels, I_{sw} further decreases with an inclination similar to the reference detector. The origin of the saddle point in both PWB curves is not clear at this point. Both curves were recorded in separate cool-downs, which is why attributing the recorded shape to a measurement artifact seems unlikely. For all curves, the steeper slope of $I_{sw}(P_{opt})$ coincides with an incipient decrease of the retrapping current I_r.

By comparing the optical power at which the switching current has dropped to 90 % of its inital value, the optical coupling efficiency of "PWB 7" ("PWB 6") can be estimated to be ≈50 % (≈20 %). This is, however, just a rough estimation of the coupling efficiency. By taking into account only the power fraction hitting the active area, the measured reference curve is shifted to smaller P_{opt} by over six orders of magnitude. A factor of two (five) could easily be within the margin of error as a result of, for example, small uncertainties in the fiber-to-detector distance or the lateral shift of the active area compared to the center of the beam. Another factor which influences the calculated optical coupling efficiency of the bonds are the different slopes of the $I_{sw}(P_{opt})$ curves. The reference chip is additionally heated by the optical power hitting the substrate around the active area. The drop in I_r of the reference curve in Fig. 6.11c is therefore observed at a much lower power than the one of the PWB samples. In fact, I_r of the reference sample starts to decrease at a similar optical power as I_{sw}. Therefore, the effects of a smaller cross-section and heating, both leading to a drop in I_{sw}, overlap in case of the reference sample. This is not the case for the PWB samples for which the heating comes into play only at much higher optical powers, as visualized by vertical dotted lines in Fig. 6.11b. As a result, the slope of the blue I_{sw} curve in Fig. 6.11b is steeper than the ones of the PWB samples at the same optical power and the optical coupling efficiency obtained by simply comparing P_{opt}, for which $I_{sw}(P_{opt}) = 0.9 I_{sw}(0\,\mathrm{W})$, is lower than the actual value.

6.3 Summary

In this chapter, the integration of 3D polymer structures with SNSPDs was investigated to enhance the optical coupling, split into free-space and fiber coupling.

Section 6.1 explored a free-space coupling method of RF-SNSPD arrays using microlenses printed on top of the active detector area. These lenses increase the effective coupling area of the detector from the small area covered by the nanowire to the much larger collection area of the lens. With a 60 μm-diameter lens and a 15 μm short nanowire covering an area of

$4.0\,\mu m \times 2.1\,\mu m$, a photon-count rate increase of factor 245 has been measured compared to an adjacent pixel without a lens. This value is reasonably in line with the lens gain calculated as the ratio between lens and detector area. Using these lenses, the small fill factor of RF-SNSPD arrays, arising from the integration with light-insensitive resonators and the limited nanowire length in order to minimize influence $L_k(I_b)$, can be enhanced and the main shortcoming of RF-SNSPDs, thus, partly overcome.

In Section 6.2, the feasibility of photonic wire bonds (PWB) to efficiently couple photons from optical fibers to SNSPDs was investigated. The bonds proved to be mechanically stable and to survive a total of four cool-downs to 4.2 K. Furthermore, leaving the resist volume in the direct vicinity of the detector unexposed has shown to prevent the degradation of the nanowire, previously encountered for nanowires under direct illumination by the high-power laser. However, the optical coupling efficiency could not be reliably determined with the methods presented here. The fact that $CR(I_b)$ was not saturated at 4.2 K did not allow for a direct, straightforward determination of η_{oc}. Calibrating the optical power using the same detector with normal photon incidence as a reference was utilized to circumvent this problem. However, the polymer material on top of the nanowire presumably caused an increase in intrinsic efficiency η_{int} compared to the reference due to thermal decoupling from the cooling environment, thereby preventing the accurate extraction of η_{oc} of the bond. This increase in intrinsic detection efficiency is actually advantageous for the detector operation but an unwanted effect for the purpose of calibration. Apart from using the nanowire as an SNSPD, another method based on the photon-induced reduction in the effective cross-section was tested by recording the switching current dependent on the incident optical power. Comparing the power level at which I_{sw} has decreased to $0.9 I_{sw}(P_{opt} = 0\,W)$ to a reference measurement, coupling efficiencies of 50 % and 20 % for the two measured PWB-coupled detectors could be estimated. For a more precise measurement of η_{oc} of the PWBs, a lower operating temperature is necessary which would allow for a higher, determinable intrinsic detection efficiency of the detector and, thus, the possibility to determine factors η_{oc} and η_{int} separately. Furthermore, the spectral bandwidth achievable with PWB-coupled SNSPDs is yet to be determined.[3] Nevertheless, the presented work shows a promising, flexible coupling scheme for SNSPDs, which could also be extended to chip-to-detector coupling for the multi-chip integration of SNSPDs with photonic circuits.

[3] In principle, the bandwidth is limited by absorption in the polymer, defining a window from $0.5\,\mu m$ [265] to around $5.0\,\mu m$ wavelength, in which the polymer is transparent.

7 Nanowire characterization with M-KING

The microwave kinetic-inductance nanowire galvanometer (M-KING) was presented by Doerner et al. as a multiplexable current sensor for cryogenic detectors [33]. The M-KING utilizes the shift in resonance frequency caused by a change in kinetic inductance when a low-frequency current is injected. Here, we use this device as a nanowire characterization method.
The ALD-NbN film was deposited by M. Ziegler and V. Ripka at Leibniz-IPHT. E. Knehr fabricated the device based on a design from Ref. [190], measured the sample with contribution from A. Kuzmin and analyzed the results.

The ratio of the switching current to the depairing current is one of the most important properties of a superconducting nanowire when evaluating its suitability as an SNSPD. A high ratio I_{sw}/I_{dep} enables biasing the detector closer to I_{dep} which, as described in Sections 2.2.2 and 2.3.1, facilitates photon-detection and leads to a longer cut-off wavelength and smaller timing jitter [8]. Acquiring accurate values of I_{dep} is therefore important for the evaluation and optimization of the material composition and patterning process of the detector.

The depairing current I_{dep} is commonly calculated from parameters determined by DC and magnetic-field measurements (see Section 3.2). From the slope of the temperature-dependent second critical magnetic field dB_{c2}/dT, the diffusion coefficient D can be calculated (see Eq. 3.3). Together with ρ_n and T_c, the depairing current over the whole temperature range $0 \leq T \leq T_c$ can then be obtained from the model of Kupriyanov and Lukichev [218] (see also Eqs. 3.5–3.7).

Alternatively, from the work of Clem and Kogan [226], the depairing current I_{dep} can directly be inferred from the current-dependent kinetic inductance $L_k(I_b)$. The latter could be obtained by resonance measurements of a suitable superconducting structure. This method enables the acquisition of $I_{dep}(T)$ dependent on the temperature and was demonstrated by Frasca *et al.* [81] who showed resonance measurements of a CPW-integrated nanowire. By fitting the phase response of the sample recorded at various bias currents, dependence $f_r(I_b)$ and, thus, the relative dependence $L_k(I_b)/L_{k,0}$, where $L_{k,0} = L_k(I_b = 0)$, was acquired. From Ref. [226], the only fitting parameter to the latter dependence is the depairing current I_{dep}.

In this chapter, a very similar method is presented, which does, however, have several advantages over the specific method from Ref. [81]. Instead of measuring the self-resonance of a CPW-integrated nanowire, the change in L_k is read out using a nanowire embedded into a lumped-element resonator with the possibility to inject a DC bias current through the nanowire. This device, the microwave kinetic-inductance nanowire galvanometer (M-KING), was originally designed as multiplexable current sensor to read out cryogenic detectors [33]. The device layout is described in Section 7.1, before going into detail of the method to extract I_{dep} of the nanowire in Sections 7.2 and 7.3. Apart from determining I_{dep}, the M-KING might also be used to extract further information on the microstructure of the nanowire. This is explored in Section 7.4 by evaluating the noise current of the device.

7.1 Layout and fabrication

The device layout is adapted from Ref. [190] with minor modifications. It consists of a parallel LC resonator with an interdigitated capacitance C and a nanowire inductance L_k, in series with an inductance L_c which is connected to the transmission line. In contrast to the capacitively coupled resonator of an RF-SNSPD from Chapter 5, the inductive coupling allows for an additional DC current through the nanowire, which can be fed by a separate DC port. Moreover, by using the nanowire as the inductive part of the resonator, the M-KING is designed in a way to maximize the current-dependent shift in resonance frequency. The magnitude of this shift largely defines the current sensitivity of the device. There are two constraints regarding the design of inductance L_c with respect to the nanowire (L_k): (i) it has to be defined by a stripe much wider than the nanowire so that L_c is essentially independent from the DC bias in the current range up to the switching current of the nanowire, and (ii) $L_c \gg L_k$ in order that there is a strong dependence of the resonator's total inductance $L = L_c L_k/(L_c + L_k)$ on $L_k(I_b)$.

An SEM micrograph of the device is shown in Fig. 7.1. Inductance L_c is defined by a meandered, 1.8 µm wide wire with a total length of 2.2 mm. Capacitance C is split in two halves (C_h) with 42 fingers in total and 1.2 µm wide gaps in between. Each finger is 0.8 µm wide and 100 µm long. A 120 nm wide and 10 µm long nanowire defines L_k. Because the meandered stripe defining L_c is about 15 times wider than the nanowire, L_c is assumed to be current-independent and the current-induced change in resonance frequency of the device is predominantly determined by $L_k(I_b)$. The DC current flows from the DC port shown at the top of Fig. 7.1 through the resonator to the ground plane, serving as the return path.

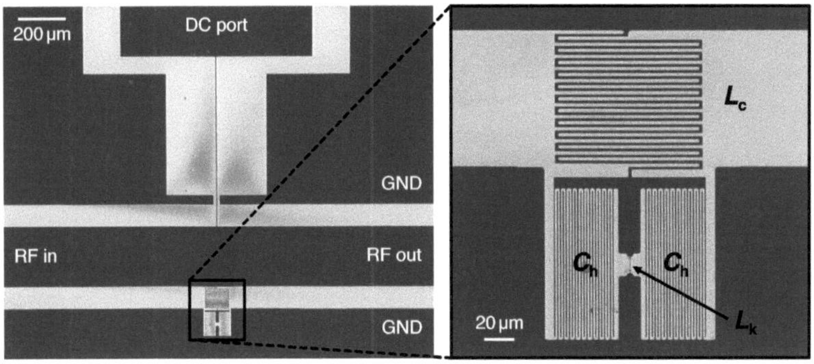

Figure 7.1: SEM micrograph of the 3-port device with an RF transmission line ("RF in" to "RF out") and a DC input port. The enlarged view shows the actual resonator which consists of a meandered inductor L_c in series with a parallel circuit of the nanowire (L_k) and a split interdigitated capacitor C. Dark areas show the NbN layer. The nanowire is highlighted by the black arrow.

The single-layer device is fabricated in accordance to the process described in Section 4.1.2. A 5 nm thick NbN layer is deposited by ALD on an R-plane sapphire substrate. The patterning is divided in three steps. First, the nanowire as the smallest structure is defined together with alignment marks in a negative-tone PMMA process. After development and reapplying of resist, both the coarse (CPW, DC port) and midsize structures (L_c, C_h) are defined in positive processes. After development, the sample is etched using Ar-ion milling.

7.2 Current-dependent microwave properties

The frequency-dependent transmission $S_{21}(f)$ on the M-KING's transmission line is measured with a vector network analyzer (VNA) using the setup shown in Fig. 7.2. The device is connected to RF and DC lines, and cooled down to 4.2 K in a LHe dewar. The RF signal from the VNA is attenuated by 40 dB at room temperature and an additional 30 dB at 4.2 K. DC drifts are prevented by DC blocks. After passing through the M-KING, the signal is amplified by 34 dB at room temperature and read out by the VNA. For noise measurements (see Section 7.4), another amplifier at 4.2 K with an amplification of 40 dB and a noise temperature of $\approx$6 K is used (see dashed components in Fig. 7.2). In this configuration, a 3 dB-attenuator is placed between both amplifiers in order to avoid reflections. The bias current through the nanowire is set via a separate DC path using a DC current source and low-pass filters with an edge frequency of $\approx$72 Hz.

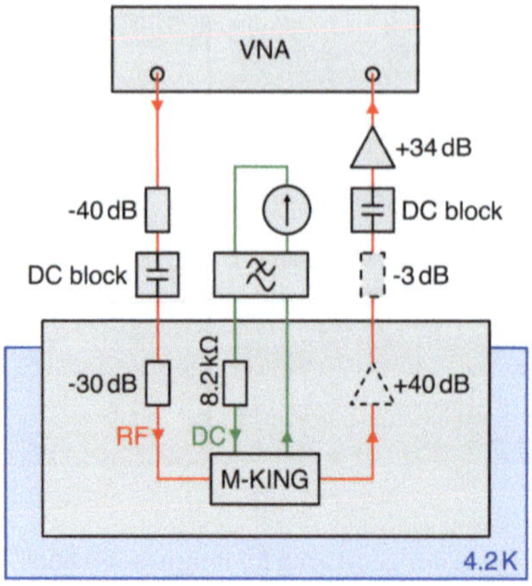

Figure 7.2: Experimental setup to measure the current-dependent microwave properties of the device. The transmission S_{21} is measured dependent on frequency and time using the VNA (red path). The DC current through the nanowire can be set using a separate, low-pass-filtered DC line (green path). For the measurements in Sections 7.2 and 7.3, the signal was only amplified at room temperature. For the noise measurements in Section 7.4, a cryogenic amplifier and 3 dB attenuator were added to the readout path.

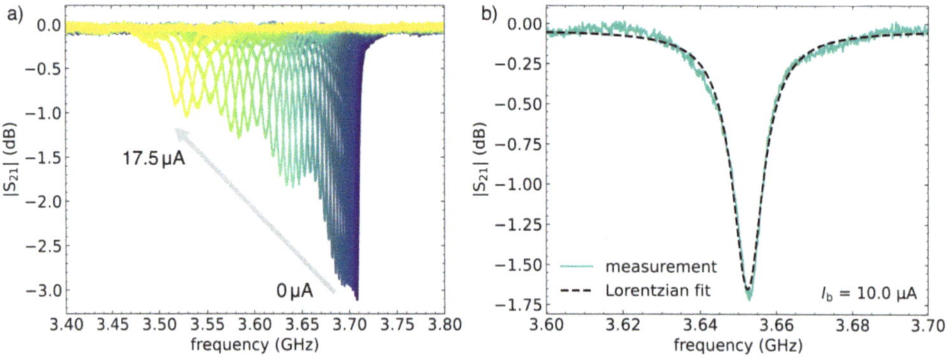

Figure 7.3: Current-dependent transmission measurements of the M-KING. (a) Magnitude of S_{21} on the transmission line for applied DC currents from 0 µA to 18 µA in steps of 0.5 µA. (b) Exemplary $|S_{21}(f)|$ curve for $I_b = 10\,\mu A$ with a Lorentzian fit (dashed line) according to Eq. 7.1. From the fit, the resonance frequency is determined to be 3.652 GHz.

The VNA's power level was set to −18 dBm, which was – after 70 dB attenuation – low enough to avoid any influence of the RF probe signal on the nanowire's inductance. This was ruled out by confirming that slight variations of the power level did not shift the resonance frequency of the M-KING. The transmission of the device was calibrated by setting a DC current above the nanowire's switching current, such that no resonance dip was observed.

The magnitude $|S_{21}(f)|$ of the transmission is shown in Fig 7.3a for a DC current ranging from 0 µA to 18 µA, with $I_{sw} = 17.8\,\mu A$. Without an applied DC current, the resonance is observed at $f_r = 3.708\,GHz$. By increasing the DC current through the nanowire, the Cooper-pair density n_s decreases (see Fig. 6 in Ref. [226]) and the penetration depth λ_l increases. This goes along with an increase in kinetic inductance L_k (see also Eq. 2.4), which is seen as a shift in resonance to lower frequencies. For a DC current $I_b = 17.5\,\mu A = 0.98 I_{sw}$ just below the switching current, the resonance has shifted to $f_r = 3.518\,GHz$. For a higher current, the resonance completely vanishes, as seen from the yellow curve in Fig. 7.3a.

The shift in resonance goes along with a smaller and wider resonance dip, thus a reduced quality factor. This is expected from the decrease in n_s (increase in quasiparticle density) at higher currents. Another observation are the slight oscillations of the resonance depth over the applied current range, which should, however, not be detrimental to the sole extraction of $f_r(I_b)$.

Precise values of f_r were extracted from the experimental data by fitting each resonance curve by a Lorentzian function

$$|S_{21}(f)| = A + \frac{|S_{21,\min}| \cdot \Delta f^2}{4\left(f - f_r\right)^2 + \Delta f^2} = A + \frac{|S_{21,\min}|}{1 + 4Q^2 \cdot \left(\frac{f-f_r}{f_r}\right)^2}, \tag{7.1}$$

with the background transmission magnitude A (close to 0 dB after calibration), resonance depth $|S_{21,\min}|$, FWHM Δf, and quality factor Q. An exemplary fit to the experimental data is shown in Fig. 7.3b for a bias current $I_b = 10\,\mu A$. The standard error of the fit in terms of f_r amounts to 14 kHz in this particular example. Compared to Ref. [81], the M-KING exhibits more pronounced resonances in the transmission magnitude, enabling reliable fits and a precise extraction of $f_r(I_b)$. The same applies to the phase response of the sample which was, however, not utilized here.

In theory, many parameters including quality factors and impedances of the resonator could be extracted. For the purpose of determining the depairing current, however, only the relative change in resonance frequency $f_r(I_b)$ with the bias current is of importance. With

$$f_r = \frac{1}{2\pi\sqrt{LC}}, \quad L = \frac{L_c L_k}{L_c + L_k}, \tag{7.2}$$

the kinetic inductance change is expressed as a function of the resonance frequency by

$$\frac{L_k(I_b)}{L_{k,0}} = \frac{L_c + L_k(I_b)}{L_c + L_{k,0}} \left[\frac{f_{r,0}}{f_r(I_b)}\right]^2 \approx \left[\frac{f_{r,0}}{f_r(I_b)}\right]^2. \tag{7.3}$$

Table 7.1: Parameters n_{fr} (fast relaxation), n_{sr} (slow relaxation), and y_0 in Eqs. 7.4 and 7.5 dependent on the relative temperature t according to Ref. [226].

$t = T/T_c$	0	0.1	0.2	0.3	0.4	0.5	0.6	0.7	0.8	0.9	$\to 1$
n_{fr}	2.21	2.21	2.27	2.30	2.28	2.25	2.22	2.18	2.16	2.13	2.11
n_{sr}	2.48	2.47	2.45	2.46	2.50	2.50	2.50	2.50	2.50	2.50	2.50
y_0	1.451	1.448	1.422	1.412	1.417	1.432	1.448	1.463	1.477	1.490	1.500

The latter approximation is reasonable because $L_c/L_k \approx 15$ which, with an expected maximum change in kinetic inductance of ≈20 %, results in an error of 1.3 % at most. As stated above, L_c is assumed to be current-independent up to the switching current of the nanowire because it is defined by a structure much wider than the nanowire.

7.3 Determination of the depairing current

The theoretical dependence of the kinetic inductance of superconducting nanostripes on the current relative to the depairing current was derived by Clem and Kogan in Ref. [226]. Two limits are considered, depending on the relative time scales of the characteristic relaxation time $\tau_{|\Delta|}$ (in the publication "τ_s") of the superconductor's order parameter and the experimental time scale τ_{expt}. If changes of the current and Cooper-pair velocity happen on a time scale τ_{expt} much longer than the relaxation time $\tau_{|\Delta|}$, the fast-relaxation (or slow-experiment) limit is applicable [226]. For $\tau_{expt} \ll \tau_{|\Delta|}$, on the other hand, the slow-relaxation (or fast-experiment) limit holds. In our case, $\tau_{|\Delta|} = \hbar/k_B T_c \approx 1\,\mathrm{ps}$ (see also Section 2.2.1) and $\tau_{expt} \approx 1/f_r \approx 270\,\mathrm{ps}$, therefore it is expected that the fast-relaxation model better describes the experimental data.

According to Ref. [226] (Fig. 7), the kinetic-inductance dependence in the fast-relaxation limit can be described by

$$\frac{L_k(I_b)}{L_{k,0}} = \left[1 - \left(\frac{I_b}{I_{dep}}\right)^{n_{fr}}\right]^{-1/n_{fr}} . \tag{7.4}$$

Parameter n_{fr} is specified in Ref. [226] dependent on the relative temperature $t = T/T_c$ and is listed in Tab. 7.1. The critical temperature T_c of the sample was obtained by standard $R(T)$ measurements using the DC ports of the device and amounts to 9.55 K (0.01 $R_{25\,\mathrm{K}}$ criteria). Because of dependence $n_{fr}(t)$, the exact definition of T_c from the experimental data, given a certain transition width ΔT of $R(T)$, has an influence on Eq. 7.4. However, assuming $I_b/I_{dep} = 0.5$ and $T = 4.2\,\mathrm{K}$, a variation of T_c by 1 K would only lead to a deviation of the kinetic inductance by ≈0.2 % and is, thus, not significant. In our case at $T = 4.2\,\mathrm{K}$, $n_{fr} \approx 2.27$.

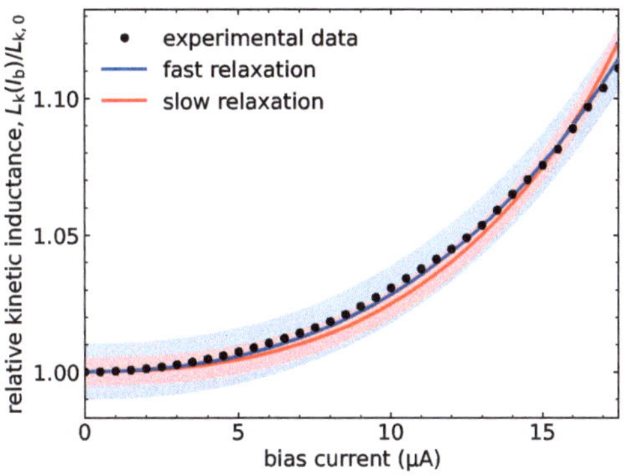

Figure 7.4: Kinetic inductance L_k over bias current relative to the value at zero bias $L_{k,0}$. The experimental data are fit by the theoretical dependence from Ref. [226] both in the fast- and slow-relaxation limit according to Eq. 7.4 and 7.5, respectively. The shaded ranges denote the accuracy of 1 % (fast relaxation) and 0.5 % (slow relaxation) as specified in Ref. [226].

Although the fast-relaxation limit was considered to be more appropriate here, for comparison the current-dependent kinetic inductance was also modeled in the slow-relaxation limit and is, according to Ref. [226] (Fig. 8), represented by

$$\frac{L_k(I_b)}{L_{k,0}} = y_0 - [y_0 - 1]\left[1 - \left(\frac{I_b}{I_{dep}}\right)^{n_{sr}}\right]^{1/n_{sr}} . \tag{7.5}$$

Similar to n_{fr}, n_{sr} and y_0 are fixed parameters at a given relative temperature and are specified in Tab. 7.1. In our case, we use approximate values of $n_{sr} = 2.5$ and $y_0 = 1.422$. For both limits, the depairing current I_{dep} is therefore the only fitting parameter.

In Fig. 7.4, the experimental data of the relative change in kinetic inductance L_k vs. the applied current through the nanowire is shown, as derived from Eq. 7.3. As expected, the kinetic inductance grows with the current the nanowire is carrying. We observe a maximum change of 11 % compared to L_k at zero bias. The maximum measurable ratio $L_k(I_b)/L_{k,0}$ depends on how close to I_{dep} the sample can be biased at and, thus, on the ratio I_{sw}/I_{dep} at the respective temperature.

Fits by Eqs. 7.4 and 7.5 to the experimental data are also plotted in Fig. 7.4. As it can be seen, the fit for the fast-relaxation limit (blue curve) represents the measured data better than the slow-relaxation limit. All data points lie well within the margin of accuracy as specified in Ref. [226]. Qualitatively similar to Fig. 3c in Ref. [81], the fit slightly deviates from the measured curve (visible here in Fig. 7.4 in the current range (5.0 to 11.0) μA). For comparison, Frasca *et al.*

also carried out numerical simulations which fit the experimental results slightly better than the approximation curves. To extract the depairing current, however, the fast-relaxation approximation from Ref. [226] should suffice.

From both fits, the depairing current of the investigated nanowire is determined to be 34.28 μA (fast relaxation) and 21.97 μA (slow relaxation), respectively. Disregarding the latter for the reasons stated above, the extracted I_{dep} value corresponds to a ratio $I_{sw}/I_{dep} = 0.52$. For comparison, we also determined the current ratio of the same sample via the Kupriyanov-Lukichev model (see Eq. 3.5). Measurements of ρ_n and T_c together with the diffusion coefficient $D = 0.43\,cm^2/s$ taken from previous ALD-NbN measurements (see Fig. 3.2) result in a ratio of 0.58. This value is consistent with another ALD-NbN sample presented in Fig. 3.3b (0.55) and is in reasonable agreement with the I_{sw}/I_{dep} ratio obtained in this chapter.

Although the investigated sample was a straight stripe and the influence of bend-related current crowding should be minimal, there are still several factors like film nonuniformities or patterning imperfections which can suppress I_{sw} considerably below I_{dep}. Current ratios in the range 0.5 to 0.6 are typical for NbN stripes of ≈5 nm thickness, ≈120 nm width, and at a temperature of $4.2\,K \approx (0.4 \text{ to } 0.5)T_c$ [80, 83, 87]. Notably, while observing almost the same ratio I_{sw}/I_{dep} for the same relative temperature, Frasca *et al.* measured values up to 0.7 for NbN nanowires at lower temperatures (see Fig. 4b in Ref. [81]).

7.4 Noise current

The M-KING was originally conceived as a multiplexable current sensor [33, 266]. Its noise current, the minimum detectable current change, is therefore one of the most crucial figures of merit when evaluating its performance. If not limited by extrinsic sources (bias, readout, ambient), the noise spectrum of the device could also be used to assess noise contributions intrinsic to the nanowire. Grain boundaries in polycrystalline films like NbN can be considered as a two-dimensional array of Josephson-coupled weak links in the current-carrying nanowire [267, 268], which might lead to fluctuations in the critical current (a phenomenon extensively investigated for Josephson-junction-based devices [269]), thus also in L_k and f_r. This and other film-intrinsic noise contributions and loss mechanisms could be investigated by resonator-integrated nanowires such as the M-KING. In the following, the method to measure the noise spectrum is shortly set out before presenting results obtained from the same device from 5 nm ALD-NbN as in previous sections.

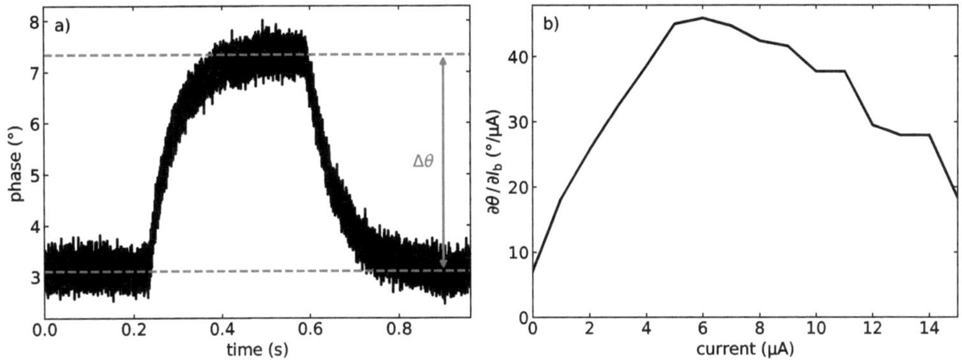

Figure 7.5: M-KING phase responsivity over bias current. (a) Exemplary time trace of the phase response at $I_b = 8\,\mu A$ and $\Delta I = 100\,nA$. The change in phase is measured to be $\Delta\theta \approx 4.24°$. (b) Phase responsivity $\partial\theta/\partial I_b$ in °/μA over bias current.

The noise current δI_{noise} was obtained from the recorded phase noise $\delta\theta_{noise}$ and the phase responsivity $\partial\theta/\partial I$ by

$$\delta I_{noise} = \delta\theta_{noise} \cdot \left(\frac{\partial\theta}{\partial I}\right)^{-1}. \tag{7.6}$$

The phase noise $\delta\theta_{noise}$ was acquired using the setup shown in Fig. 7.2 including the first-stage amplifier at 4.2 K. Time traces of the transmitted phase signal were recorded in continuous-wave mode of the VNA with $f = f_r(I_b)$. For each DC bias current in the range (0 to 15) μA = (0 to 0.84)I_{sw}, ten time traces of 960 ms length each were recorded. The spectrum was calculated as

$$\delta\theta_{noise} = \sqrt{\frac{|\mathrm{FFT}(\theta)|^2 \cdot \Delta t}{k}}, \tag{7.7}$$

with phase θ, time difference Δt between subsequent recorded data points, and number of points k.

The phase responsivity $\partial\theta/\partial I$ was approximated by measuring $\Delta\theta/\Delta I$. For this purpose, the DC current was varied by $\Delta I = 100\,nA$ at each bias point, which caused a change in recorded phase. Figure 7.5a shows an exemplary time trace of this measurement at $I_b = 8\,\mu A + \Delta I$. Current ΔI is injected at $t \approx 0.25$ s and removed again at $t \approx 0.60$ s, leading to a change in phase of $\Delta\theta \approx 4.24°$.

The resulting phase responsivity over bias current is shown in Fig. 7.5b. As it is seen, the responsivity reaches a maximum of ≈45 °/μA in the bias range (5.0 to 7.0) μA or (0.3 to 0.4)I_{sw}. As illustrated in Ref. [33], the maximum responsivity of the device in this current range is the

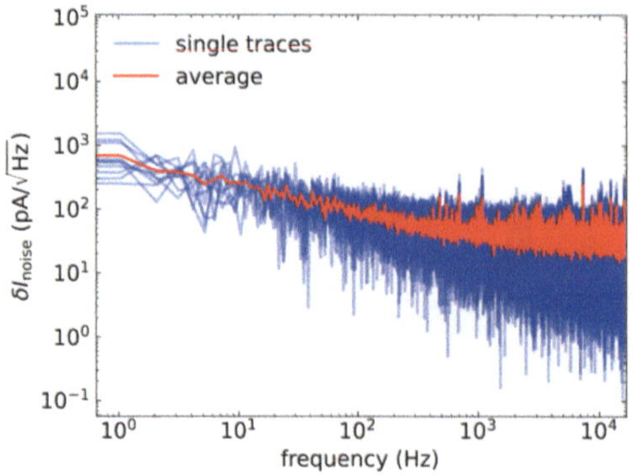

Figure 7.6: Noise-current spectrum $\delta I_{noise}(f)$ recorded at $I_b = 7.0\,\mu A$ with ten individual traces (blue) and the average spectrum (red).

result of two factors. On the one side, the change in resonance frequency is the highest at currents close to I_{sw} (see the slope in Fig. 7.4). On the other side, the quality factor decreases with higher bias currents (see Fig. 7.3), thereby reducing the influence of the bias current on the phase response. Similarly, the minimum noise current is a trade-off between high responsivity and low phase noise. The latter might be, for example, influenced by additional fluctuations at currents close to the switching current.

The lowest noise current δI_{noise} was found at $I_b = 7.0\,\mu A = 0.39 I_{sw} = 0.20 I_{dep}$ and is shown in Fig. 7.6. Ten individually recorded spectra as well as the average spectrum are plotted. Below $\approx$500 Hz, the noise current exhibits a $1/f$ dependence while a white noise level of 35 pA/$\sqrt{Hz}$ is observed at higher frequencies.

These characteristics closely resemble the ones recorded before with an M-KING from sputtered NbN of the same film thickness of 5 nm, at the same operating temperature, and using virtually the same measurement setup (see Fig. 6 in Ref. [266]). This raises the question whether the recorded noise level might be defined by, e.g., the bias and readout electronics or other sources, such as generation-recombination noise (thermally excited quasiparticles) or two-level system noise [270]. While the latter two are considered to be irrelevant for the given circumstances in Ref. [33], the frequency-independent white noise level is most likely limited by readout noise of the particular setup here. In a subsequent noise measurement (using an M-KING from sputtered NbN), the VNA was replaced by a signal generator and a homodyne receiver with a 24-bit ADC. With this setup, a white noise level of only few pA/$\sqrt{Hz}$ was recorded, well below the (35 to 40) pA/$\sqrt{Hz}$ measured here and in Ref. [266].

The observed low-frequency noise, however, was comparable in all of the measurements independent of the setup (with variations much smaller than the discrepancy in the white noise level), and might be indicative of nanowire-inherent noise due to its microstructure. More extensive measurements of several samples at various operating temperatures together with morphological investigations of the film structures are necessary to better classify the results and assess the possibility of using the M-KING as a superconducting nanowire characterization tool.

7.5 Summary

This chapter presented the microwave kinetic-inductance nanowire galvanometer (M-KING) as a device to characterize superconducting nanowires and, in particular, to determine their switching to depairing current ratio, a metric of high importance for SNSPDs. The M-KING was previously demonstrated as a current sensor which takes advantage of the current-dependent change in kinetic inductance of the nanowire [33, 266].

Using the bias-dependent resonance frequency, dependence $L_k(I_b)/L_{k,0}$ can be extracted. This nonlinear curve is then fit by theoretical models in the fast- and slow-relaxation limit from Clem and Kogan [226] with I_{dep} as the only fitting parameter.

Compared to the method shown by Frasca *et al.* [81], the presented implementation has the advantage of (i) a simpler microwave measurement setup, (ii) more pronounced resonances in the device's magnitude response enabling a more reliable fit to extract $f_r(I_b)$, (iii) a flexible layout to set an appropriate resonance frequency range for various nanowire geometries, and (iv) the possibility to also measure the microwave transmission of the device without any resonance dip by switching the nanowire to the normal state, e.g., for calibration purposes without a separate reference sample.

A proof-of-principle device, which was fabricated from 5 nm thick ALD-NbN with a nanowire width of 120 nm, exhibits a ratio $I_{sw}/I_{dep} = 0.52$ at 4.2 K. This value is in the expected range 0.5 to 0.6 of NbN nanowires at this operating temperature, and resembles the ratios extracted by the conventional method using DC transport measurements (see Section 3.2).

Besides the current ratio, the M-KING might also be used to evaluate the nanowire quality by noise spectrum measurements, possibly providing insights into the film morphology.

8 Conclusions and outlook

In this work, advances in three areas of SNSPD development have been made, all aiming toward the extension of few-pixel devices to larger arrays: employing atomic layer deposition for uniform superconducting NbN thin films over large areas, investigations of the frequency-domain multiplexing scheme for SNSPDs to quantify and overcome some of its limitations, and the integration of SNSPDs with free-form polymers to enhance both free-space and fiber optical coupling.

The importance of spatial uniformity in geometry and material parameters for SNSPDs is twofold. Superconducting thin films uniform on the nano- and micrometer-scale enable to bias the detectors near the optimum along the whole nanowire, thereby extending the photon sensitivity, reducing dark counts, and lowering the timing jitter. Besides, a small property spread on the millimeter-scale allows for minimal pixel-to-pixel variation which is crucial for the fabrication of large-area detector arrays and the fabrication yield in general. For this purpose, an existing atomic layer deposition process as an alternative to commonly used sputtering for the growth of NbN films was investigated specifically in view of the SNSPD requirements. Basic properties of (3 to 21) nm thick films indicate that ALD-NbN is a "worse" superconductor than sputtered NbN, though actually beneficial for the photon-detection mechanism of SNSPDs. The small electron-diffusion coefficient below $0.5\,\mathrm{cm^2/s}$, moderate critical temperature of $\approx 10\,\mathrm{K}$ (proportional to the energy gap), and high residual resistivity above (300 to 400) $\mu\Omega\,\mathrm{cm}$ (meaning low density of electron states) for 5 nm thick ALD-NbN are all indications of a dirty superconductor. These properties facilitate photon detection because the absorbed photon energy is confined in a smaller volume and generates a larger fraction of excited quasiparticles, thereby causing a faster thermalization and a stronger suppression of the superconducting state within the hotspot. In line with the tendency of these results, the absolute switching current density of an ALD-NbN stripe of $10\,\mathrm{nm} \times 100\,\mathrm{nm}$ cross-section was considerably lower at $9\,\mathrm{MA/cm^2}$ than values on sputtered NbN stripes, but the switching to depairing ratio was virtually the same, reaching values close to 0.6.

In an effort to optimize the ALD process, the plasma-to-substrate distance during deposition was varied. It was found that the stronger plasma interaction with the substrate surface of samples closer to the plasma source led to gradually degrading film properties, which suggested that lowering the plasma power might actually be beneficial. Because at some point the reactivity is too low, an optimum plasma power has yet to be found.

Investigations of the property spread of ALD-NbN across the substrate table showed that the superconducting properties were near-constant over a diameter of 50 mm. While the film thickness and surface roughness were unchanged also further from the table center, XRD measurements disclosed a change in film crystallinity, in particular a shrinking lattice constant, which most likely caused the observed degradation in critical temperature and switching current density, among others. The dimensions of the uniform area coincided closely with the diameter of the plasma source, indicating the strong influence of the plasma species in the downstream region of the plasma source on the film growth. A larger plasma source could therefore extend the film uniformity.

First SNSPDs fabricated from ALD-NbN films exhibited a single-photon response and saturating photon-count rates for photons in the visible spectrum, equivalent to near-unity intrinsic detection efficiency. From spectral measurements, a cut-off wavelength of about 1.25 µm was obtained for a 5 nm thick detector biased at $0.9I_{\mathrm{sw}}$ and operated at 4.2 K. Compared to an SNSPD twice as thick, this is an increase in cut-off wavelength by a factor of 1.6 at the same relative bias current, as qualitatively expected from the smaller vortex-entry barrier for thinner detectors. The suitability of ALD-NbN for SNSPD, especially for large-area devices with high fabrication yield, have since been confirmed by other groups [225, 271]. While NbN-based SNSPDs are likely not appropriate for the detection of mid-infrared photons, NbN is the material-of-choice for fast detectors in the visible and near-infrared range including the telecommunication wavelengths. Besides NbN, ALD might also be adopted for SNSPDs from other materials like NbSi [272] or NbC [273, 274] in the future.

The second area of detector development, the bias and readout of detector arrays, was targeted by further investigations of the frequency-domain multiplexing scheme demonstrated for SNSPDs ("RF-SNSPD") in previous studies. Herein, the focus was on the various implications of the oscillating bias current. As measurements of the bias-dependent dark-count rate showed, it eliminates latching and leads to a stable detector operation without the need of an active reset. This is important if the detector operation might otherwise be frequently impeded, e.g., by backscattered photons in laser-ranging applications. Electrothermal simulations illustrated how a photon-induced hotspot collapses with the amplitude of the microwave bias. In contrast to a simulated DC-biased nanowire, the RF bias reliably reset the detector after each detection event, even for reset times of only 0.4 ns. Another implication of the microwave bias is the reduced duty cycle of the detector. Analyzing detection events dependent on the bias phase revealed that the instantaneous intrinsic detection efficiency follows the microwave bias current and leads to a gated detection with a frequency twice the resonance frequency. This was exploited for a gated detection scheme suitable for synchronous applications where the photon-arrival time spans are known. Synchronizing the detector to an amplitude-modulated VCSEL diode at 3.8 GHz showed a photon-count rate oscillating with a modulation index of 0.71, depending on the relative phase between the microwave bias and the optical signal. The results were in good agreement with a simple model of the operation conditions and demonstrated that the signal-to-noise ratio can be maximized and the stray light influence minimized by this synchronized detection scheme. For random photon arrivals, on the other hand, the oscillating intrinsic detection efficiency leads to a reduced average efficiency (in our case only 30 %) which could only be alleviated using detectors with a broad saturation over bias. In this case, however, the detector's duty cycle could be tuned freely between gated and quasi-free-running mode.

A readout method using correlation filters on the common signal of a five-pixel detector array was tested to trace back photon events to the triggering pixel, thereby extracting spatial information on the detection event. In a proof-of-pinciple demonstration, the signal was post-processed using pulse templates of each pixel, which showed the reliable pixel attribution even in the case of temporally overlapping events. Photon statistics recorded simultaneously using two detectors replicated the expected Poisson distribution of coherent photons from a laser diode. In the future, the matched-filter readout could be implemented in real time together with the generation of the probe-tone comb for biasing on an FPGA-based platform. In general, scaling SNSPD arrays beyond the current record in kilopixel range will most likely require the combination of multiple multiplexing architectures, of which frequency-domain multiplexing could be one part.

One of the main shortcomings of SNSPDs in general and RF-SNSPDs in particular is the small photon-sensitive area which impedes an efficient coupling of photons from the detection system input. In order to enhance the coupling efficiency, SNSPDs were integrated with free-form polymers structures – microlenses and photonic wire bonds – which were already widely in use for silicon photonics at room temperature. These optical elements were fabricated by direct laser writing using two-photon absorption.

An RF-SNSPD array comprising pixels with and without a microlens on top of the nanowire was used to study the increase in free-space optical coupling. A 60 µm-diameter lens on top of the nanowire covering an area of 4.0 µm × 2.1 µm led to a gain in photon-count rate of about 245, which is in line with the area increase from the detector to the collection area of the lens. A microlens-integrated detector array could therefore considerably increase the effective fill factor of RF-SNSPDs.

Photonic wire bonds as a method to efficiently couple photons from fibers to SNSPDs were fabricated by the same technology to connect an eight-channel fiber array to a DC-biased

SNSPD array. The bonds proved to be mechanically stable over four conducted cryo-cycles to 4.2 K. Moreover, leaving a small resist volume in the direct vicinity of the nanowire unexposed during the lithography prevented nanowire degradation. The optical coupling efficiency from the single-mode fiber to the detector could not be directly inferred because of the low intrinsic efficiency of the detectors at 4.2 K and 1550 nm wavelength. Instead, the efficiency was tried to be obtained by measuring the nanowire's switching current dependent on the photon rate. The simultaneous absorption of multiple photons results in a reduced superconducting cross-section. Approximate coupling efficiencies of two photonic wire bonds deduced with this method amounted to 50 % and 20 %. A direct determination at lower temperatures will be necessary for a more precise determination of the coupling efficiency achievable by photonic wire bonds at cryogenic temperatures.

In the future, photonic-wire-bond coupling of SNSPDs should be directly extendable to couple photons from on-chip waveguides, which would allow for a flexible multi-chip integration of SNSPDs with photonic circuits. Instead of the normal-incidence geometry shown here, a configuration with the photonic wire bond laying along the nanowire would be beneficial because of the much longer absorption length achievable even for short nanowires. Moreover, because of the 3D arrangement, a straight nanowire along a waveguide without any bend would be feasible which is not possible with planar waveguides.

Complementing the development on scaling SNSPDs, a method to extract the switching to depairing current ratio was presented, which takes advantage of the current-dependent kinetic inductance of a resonator-integrated nanowire. Current ratios in line with values from transport measurements were obtained. Compared to previous studies, the presented method allows for an easier measurement setup with on-device calibration.

List of Figures

List of Tables

Nomenclature

Abbreviations

ALD	atomic layer deposition
ADC	analog-to-digital converter
AFM	atomic force microscopy
BCS	Bardeen-Cooper-Schrieffer (theory of superconductivity)
CPW	coplanar waveguide
CR	count rate
CVD	chemical vapor deposition
DC	direct current
DCR	dark-count rate
DE	detection efficiency
DLW	direct laser writing
EBL	electron-beam lithography
FDM	frequency-division multiplexing
FFT	fast Fourier transform
FWHM	full width at half maximum
GPC	growth per cycle

ICP	inductively coupled plasma
IF	intermediate frequency
IPA	isopropyl alcohol
KIB	kinetic-inductance bolometer
KL	Kupriyanov-Lukichev (model)
LHe	liquid helium
LO	local oscillator
MCR	maximum count rate
MFD	mode-field diameter
MIBK	methyl isobutyl ketone
MKID	microwave kinetic-inductance detector
M-KING	microwave kinetic-inductance nanowire galvanometer
PCB	printed circuit board
PCR	photon-count rate
PEALD	plasma-enhanced ALD
PMMA	poly (methyl methacrylate)
PR	photon rate
PWB	photonic wire bond
RF	radio frequency
RF-SNSPD	radio-frequency SNSPD
RMS	root mean square
SDE	system detection efficiency

SEM	scanning electron microscopy
SIMS	secondary ion mass spectroscopy
SMF	single-mode fiber
SNR	signal-to-noise ratio
SNSPD	superconducting nanowire single-photon detector
TBTDEN	(*tert*-butylimido)-tris(diethylamino)-niobium
TES	transition-edge sensor
VAP	vortex-antivortex pair
VCSEL	vertical-cavity surface-emitting laser
VNA	vector network analyzer
XRD	x-ray diffraction

Symbols

β_0	ratio between $\Delta(0)$ and $k_B T_c$
Γ	vortex-hopping rate
γ	mean photon rate
Δ	superconducting energy gap
Δf	resonance width
δf	detuning frequency
ΔI_b	transition width of current-dependent photon-count rate
ΔT	superconducting transition width
δI_{noise}	noise current
$\delta\theta_{noise}$	phase noise

δt	inter-pulse time
η_{abs}	absorption efficiency
η_{int}	intrinsic detection efficiency
$\overline{\eta}_{\mathrm{int}}$	average intrinsic detection efficiency
η_{oc}	optical coupling efficiency
2Θ	angle between incident and diffracted beam in XRD
θ	phase
Λ	Pearl length (effective magnetic penetration depth in thin films)
λ	photon wavelength
λ_{c}	cut-off wavelength
λ_{l}	London penetration depth
μ_0	vacuum permeability
v_0	characteristic velocity of hotspot growth
ξ	coherence length
ξ_0	BCS coherence length
ξ_{GL}	Ginzburg-Landau coherence length
ρ_{n}	residual resistivity
τ_{d}	decay time
$\tau_{\lvert\Delta\rvert}$	relaxation time of the superconducting order parameter
τ_{fall}	fall time
τ_{r}	reset time
τ_{rep}	laser repetition period

τ_{rise}	rise time
Φ_0	magnetic flux quantum
ψ	Stekly parameter
ω	angular frequency; incident beam angle in XRD
A_{c}	collection area
A_{d}	detector area
B_{c2}	second critical magnetic field
C	capacitance
C_{e}	electron heat capacity
C_{ph}	phonon heat capacity
D	electron diffusion coefficient
d	thickness
$\mathrm{d}F$	energy barrier for vortex entry
$\mathrm{d}I$	excess current
e	elementary charge
E_{ph}	photon energy
f	frequency; focal length
f_{g}	gating frequency
f_{opt}	optical modulation frequency
f_{r}	resonance frequency
G_{d}	lens gain
H	magnetic-field strength

h	Planck constant; height
$\hbar$	reduced Planck constant
I	current
I_{b}	bias current
I_{dep}	depairing current
I_{det}	detection current
I_{n}	current through the nanowire
I_{r}	retrapping current
I_{rf}	RF current
I_{sw}	switching current
j_{dep}	depairing current density
j_{sw}	switching current density
k_{B}	Boltzmann constant
L	inductance
l	length
l_0	electron mean free path
l_{hs}	hotspot length
L_{k}	kinetic inductance
L_{s}	surface inductance
m_{e}	electron mass
n	refractive index
$N(0)$	single-spin electron density of states at the Fermi level

n_s	Cooper-pair density
P_opt	optical power
Q	(loaded) quality factor
R	resistance
R_abs	absorption rate
R_col	collection rate
R_hs	hotspot resistance
R_l	load resistance
R_n	normal-state resistance
R_q	surface roughness (RMS)
R_sh	normal-state sheet resistance (DC)
R_s	surface resistance
s	discrete time trace
S_{21}	transmission coefficient from port 1 to port 2
$\lvert S_{21}\rvert$	magnitude of S_{21}
$\lvert S_{21,\mathrm{min}}\rvert$	resonance depth of $\lvert S_{21}\rvert$
T	temperature
t	time; relative temperature T/T_c
T_c	critical temperature
t_p	pulse template of pixel p
V	voltage
w	width; beam radius

x, y, z	geometrical coordinates
Z_s	surface impedance

Bibliography

[1] A. Migdall, S. V. Polyakov, J. Fan, and J. C. Bienfang, editors, *Single-photon generation and detection: physics and applications*, 1st edition, Experimental methods in the physical sciences (Academic Press, 2013).

[2] M. Ghioni, A. Gulinatti, I. Rech, F. Zappa, and S. Cova, "Progress in Silicon Single-Photon Avalanche Diodes", IEEE Journal of Selected Topics in Quantum Electronics **13**, 852 (2007).

[3] S. Cova, M. Ghioni, A. Lacaita, C. Samori, and F. Zappa, "Avalanche photodiodes and quenching circuits for single-photon detection", Applied Optics **35**, 1956 (1996).

[4] G. N. Gol'tsman, O. Okunev, G. Chulkova, A. Lipatov, A. Semenov, K. Smirnov, B. Voronov, A. Dzardanov, C. Williams, and R. Sobolewski, "Picosecond superconducting single-photon optical detector", Applied Physics Letters **79**, 705 (2001).

[5] A. D. Semenov, G. N. Gol'tsman, and A. A. Korneev, "Quantum detection by current carrying superconducting film", Physica C: Superconductivity **351**, 349 (2001).

[6] A. Korneev, P. Kouminov, V. Matvienko, G. Chulkova, K. Smirnov, B. Voronov, G. N. Gol'tsman, M. Currie, W. Lo, K. Wilsher, J. Zhang, W. Słysz, A. Pearlman, A. Verevkin, and R. Sobolewski, "Sensitivity and gigahertz counting performance of NbN superconducting single-photon detectors", Applied Physics Letters **84**, 5338 (2004).

[7] D. V. Reddy, R. R. Nerem, S. W. Nam, R. P. Mirin, and V. B. Verma, "Superconducting nanowire single-photon detectors with 98% system detection efficiency at 1550 nm", Optica **7**, 1649 (2020).

[8] B. Korzh, Q.-Y. Zhao, J. P. Allmaras, S. Frasca, T. M. Autry, E. A. Bersin, A. D. Beyer, R. M. Briggs, B. Bumble, M. Colangelo, G. M. Crouch, A. E. Dane, T. Gerrits, A. E. Lita, F. Marsili, G. Moody, C. Peña, E. Ramirez, J. D. Rezac, N. Sinclair, M. J. Stevens, A. E. Velasco, V. B. Verma, E. E. Wollman, S. Xie, D. Zhu, P. D. Hale, M. Spiropulu, K. L. Silverman, R. P. Mirin, S. W. Nam, A. G. Kozorezov, M. D. Shaw, and K. K. Berggren, "Demonstration of sub-3 ps temporal resolution with a superconducting nanowire single-photon detector", Nature Photonics **14**, 250 (2020).

[9] E. E. Wollman, V. B. Verma, A. D. Beyer, R. M. Briggs, B. Korzh, J. P. Allmaras, F. Marsili, A. E. Lita, R. P. Mirin, S. W. Nam, and M. D. Shaw, "UV superconducting nanowire single-photon detectors with high efficiency, low noise, and 4 K operating temperature", Optics Express **25**, 26792 (2017).

[10] F. Marsili, F. Bellei, F. Najafi, A. E. Dane, E. A. Dauler, R. J. Molnar, and K. K. Berggren, "Efficient Single Photon Detection from 500 nm to 5 μm Wavelength", Nano Letters **12**, 4799 (2012).

[11] V. B. Verma, B. Korzh, A. B. Walter, A. E. Lita, R. M. Briggs, M. Colangelo, Y. Zhai, E. E. Wollman, A. D. Beyer, J. P. Allmaras, B. Bumble, H. Vora, D. Zhu, E. Schmidt, K. K. Berggren, R. P. Mirin, S. W. Nam, and M. D. Shaw, "Single-photon detection in the mid-infrared up to 10 micron wavelength using tungsten silicide superconducting nanowire detectors", arXiv preprint arXiv:2012.09979 (2020).

[12] E. E. Wollman, V. B. Verma, A. B. Walter, J. Chiles, B. Korzh, J. P. Allmaras, Y. Zhai, A. E. Lita, A. N. McCaughan, E. Schmidt, S. Frasca, R. P. Mirin, S.-W. Nam, and M. D. Shaw, "Recent advances in superconducting nanowire single-photon detector technology for exoplanet transit spectroscopy in the mid-infrared", Journal of Astronomical Telescopes, Instruments, and Systems **7**, 1 (2021).

[13] L. Chen, D. Schwarzer, J. A. Lau, V. B. Verma, M. J. Stevens, F. Marsili, R. P. Mirin, S. W. Nam, and A. M. Wodtke, "Ultra-sensitive mid-infrared emission spectrometer with sub-ns temporal resolution", Optics Express **26**, 14859 (2018).

[14] L. Chen, J. A. Lau, D. Schwarzer, J. Meyer, V. B. Verma, and A. M. Wodtke, "The Sommerfeld ground-wave limit for a molecule adsorbed at a surface", Science **363**, 158 (2019).

[15] M. Shaw, F. Marsili, A. Beyer, R. Briggs, J. Allmaras, and W. H. Farr, "Superconducting nanowire single photon detectors for deep space optical communication (Conference Presentation)", Proceedings of SPIE **10096**, 100960J (2017).

[16] A. McCarthy, N. J. Krichel, N. R. Gemmell, X. Ren, M. G. Tanner, S. N. Dorenbos, V. Zwiller, R. H. Hadfield, and G. S. Buller, "Kilometer-range, high resolution depth imaging via 1560 nm wavelength single-photon detection", Optics Express **21**, 8904 (2013).

[17] H. Li, S. Chen, L. You, W. Meng, Z. Wu, Z. Zhang, K. Tang, L. Zhang, W. Zhang, X. Yang, X. Liu, Z. Wang, and X. Xie, "Superconducting nanowire single photon detector at 532 nm and demonstration in satellite laser ranging", Optics Express **24**, 3535 (2016).

[18] L. Kong, Q. Zhao, H. Wang, J. Guo, H. Lu, H. Hao, S. Guo, X. Tu, L. Zhang, X. Jia, L. Kang, X. Wu, J. Chen, and P. Wu, "Single-Detector Spectrometer Using a Superconducting Nanowire", Nano Letters **21**, 9625 (2021).

[19] Y. Hochberg, I. Charaev, S.-W. Nam, V. Verma, M. Colangelo, and K. K. Berggren, "Detecting Sub-GeV Dark Matter with Superconducting Nanowires", Physical Review Letters **123**, 151802 (2019).

[20] R. H. Hadfield, "Single-photon detectors for optical quantum information applications", Nature Photonics **3**, 696 (2009).

[21] R. H. Hadfield and G. Johansson, editors, *Superconducting Devices in Quantum Optics*, 1st edition, Quantum Science and Technology (Springer, 2016).

[22] H. Takesue, S. W. Nam, Q. Zhang, R. H. Hadfield, T. Honjo, K. Tamaki, and Y. Yamamoto, "Quantum key distribution over a 40-dB channel loss using superconducting single-photon detectors", Nature Photonics **1**, 343 (2007).

[23] F. Bussières, C. Clausen, A. Tiranov, B. Korzh, V. B. Verma, S. W. Nam, F. Marsili, A. Ferrier, P. Goldner, H. Herrmann, C. Silberhorn, W. Sohler, M. Afzelius, and N. Gisin, "Quantum teleportation from a telecom-wavelength photon to a solid-state quantum memory", Nature Photonics **8**, 775 (2014).

[24] H. Takesue, S. D. Dyer, M. J. Stevens, V. Verma, R. P. Mirin, and S. W. Nam, "Quantum teleportation over 100 km of fiber using highly efficient superconducting nanowire single-photon detectors", Optica **2**, 832 (2015).

[25] R. Valivarthi, M. l. G. Puigibert, Q. Zhou, G. H. Aguilar, V. B. Verma, F. Marsili, M. D. Shaw, S. W. Nam, D. Oblak, and W. Tittel, "Quantum teleportation across a metropolitan fibre network", Nature Photonics **10**, 676 (2016).

[26] A. N. McCaughan, "Readout architectures for superconducting nanowire single photon detectors", Superconductor Science and Technology **31**, 040501 (2018).

[27] E. E. Wollman, V. B. Verma, A. E. Lita, W. H. Farr, M. D. Shaw, R. P. Mirin, and S. W. Nam, "Kilopixel array of superconducting nanowire single-photon detectors", Optics Express **27**, 35279 (2019).

[28] S. Doerner, A. Kuzmin, S. Wuensch, K. Ilin, and M. Siegel, "Operation of Superconducting Nanowire Single-Photon Detectors Embedded in Lumped-Element Resonant Circuits", IEEE Transactions on Applied Superconductivity **26**, 1 (2016).

[29] C. Hamilton, C. Burroughs, and S. Benz, "Josephson Voltage Standard - A Review", IEEE Transactions on Applied Superconductivity **7**, 3756 (1997).

[30] M. H. Devoret and R. J. Schoelkopf, "Superconducting Circuits for Quantum Information: An Outlook", Science **339**, 1169 (2013).

[31] R. L. Fagaly, "Superconducting quantum interference device instruments and applications", Review of Scientific Instruments **77**, 101101 (2006).

[32] J. Bardeen, "Two-Fluid Model of Superconductivity", Physical Review Letters **1**, 399 (1958).

[33] S. Doerner, A. Kuzmin, K. Graf, I. Charaev, S. Wuensch, and M. Siegel, "Compact microwave kinetic inductance nanowire galvanometer for cryogenic detectors at 4.2 K", Journal of Physics Communications **2**, 025016 (2018).

[34] J. Bardeen, L. N. Cooper, and J. R. Schrieffer, "Theory of Superconductivity", Physical Review **108**, 1175 (1957).

[35] B. Mühlschlegel, "Die thermodynamischen Funktionen des Supraleiters", Zeitschrift für Physik **155**, 313 (1959).

[36] U. S. Pracht, E. Heintze, C. Clauss, D. Hafner, R. Bek, D. Werner, S. Gelhorn, M. Scheffler, M. Dressel, D. Sherman, B. Gorshunov, K. S. Ilin, D. Henrich, and M. Siegel, "Electrodynamics of the Superconducting State in Ultra-Thin Films at THz Frequencies", IEEE Transactions on Terahertz Science and Technology **3**, 269 (2013).

[37] V. V. Schmidt, *The Physics of Superconductors: Introduction to Fundamentals and Applications*, edited by P. Müller and A. V. Ustinov, 1st edition (Springer, 1997).

[38] D. Y. Vodolazov, "Single-Photon Detection by a Dirty Current-Carrying Superconducting Strip Based on the Kinetic-Equation Approach", Physical Review Applied **7**, 034014 (2017).

[39] A. Engel, J. J. Renema, K. Il'in, and A. Semenov, "Detection mechanism of superconducting nanowire single-photon detectors", Superconductor Science and Technology **28**, 114003 (2015).

[40] A. Semenov, A. Engel, H.-W. Hübers, K. Il'in, and M. Siegel, "Spectral cut-off in the efficiency of the resistive state formation caused by absorption of a single-photon in current-carrying superconducting nano-strips", The European Physical Journal B - Condensed Matter and Complex Systems **47**, 495 (2005).

[41] L. N. Bulaevskii, M. J. Graf, C. D. Batista, and V. G. Kogan, "Vortex-induced dissipation in narrow current-biased thin-film superconducting strips", Physical Review B **83**, 144526 (2011).

[42] L. N. Bulaevskii, M. J. Graf, and V. G. Kogan, "Vortex-assisted photon counts and their magnetic field dependence in single-photon superconducting detectors", Physical Review B **85**, 014505 (2012).

[43] A. Engel and A. Schilling, "Numerical analysis of detection-mechanism models of superconducting nanowire single-photon detector", Journal of Applied Physics **114**, 214501 (2013).

[44] A. Engel, J. Lonsky, X. Zhang, and A. Schilling, "Detection Mechanism in SNSPD: Numerical Results of a Conceptually Simple, Yet Powerful Detection Model", IEEE Transactions on Applied Superconductivity **25**, 1 (2015).

[45] A. N. Zotova and D. Y. Vodolazov, "Photon detection by current-carrying superconducting film: A time-dependent Ginzburg-Landau approach", Physical Review B **85**, 024509 (2012).

[46] D. Y. Vodolazov, "Current dependence of the red boundary of superconducting single-photon detectors in the modified hot-spot model", Physical Review B **90**, 054515 (2014).

[47] A. N. Zotova and D. Y. Vodolazov, "Intrinsic detection efficiency of superconducting nanowire single photon detector in the modified hot spot model", Superconductor Science and Technology **27**, 125001 (2014).

[48] D. Y. Vodolazov, Y. P. Korneeva, A. V. Semenov, A. A. Korneev, and G. N. Goltsman, "Vortex-assisted mechanism of photon counting in a superconducting nanowire single-photon detector revealed by external magnetic field", Physical Review B **92**, 104503 (2015).

[49] Y. P. Korneeva, D. Y. Vodolazov, A. V. Semenov, I. N. Florya, N. Simonov, E. Baeva, A. A. Korneev, G. N. Goltsman, and T. M. Klapwijk, "Optical Single-Photon Detection in Micrometer-Scale NbN Bridges", Physical Review Applied **9**, 064037 (2018).

[50] J. K. W. Yang, A. J. Kerman, E. A. Dauler, V. Anant, K. M. Rosfjord, and K. K. Berggren, "Modeling the Electrical and Thermal Response of Superconducting Nanowire Single-Photon Detectors", IEEE Transactions on Applied Superconductivity **17**, 581 (2007).

[51] H. Bartolf, *Fluctuation Mechanisms in Superconductors*, 1st edition (Springer Spektrum, 2016).

[52] A. J. Kerman, E. A. Dauler, W. E. Keicher, J. K. W. Yang, K. K. Berggren, G. Gol'tsman, and B. Voronov, "Kinetic-inductance-limited reset time of superconducting nanowire photon counters", Applied Physics Letters **88**, 111116 (2006).

[53] A. J. Kerman, J. K. W. Yang, R. J. Molnar, E. A. Dauler, and K. K. Berggren, "Electrothermal feedback in superconducting nanowire single-photon detectors", Physical Review B **79**, 100509 (2009).

[54] F. Marsili, M. J. Stevens, A. Kozorezov, V. B. Verma, C. Lambert, J. A. Stern, R. D. Horansky, S. Dyer, S. Duff, D. P. Pappas, A. E. Lita, M. D. Shaw, R. P. Mirin, and S. W. Nam, "Hotspot relaxation dynamics in a current-carrying superconductor", Physical Review B **93**, 094518 (2016).

[55] K. S. Il'in, M. Lindgren, M. Currie, A. D. Semenov, G. N. Gol'tsman, R. Sobolewski, S. I. Cherednichenko, and E. M. Gershenzon, "Picosecond hot-electron energy relaxation in NbN superconducting photodetectors", Applied Physics Letters **76**, 2752 (2000).

[56] R. W. Heeres and V. Zwiller, "Superconducting detector dynamics studied by quantum pump-probe spectroscopy", Applied Physics Letters **101**, 112603 (2012).

[57] L. Zhang, L. You, X. Yang, J. Wu, C. Lv, Q. Guo, W. Zhang, H. Li, W. Peng, Z. Wang, and X. Xie, "Hotspot relaxation time of NbN superconducting nanowire single-photon detectors on various substrates", Scientific Reports **8**, 1486 (2018).

[58] A. J. Annunziata, O. Quaranta, D. F. Santavicca, A. Casaburi, L. Frunzio, M. Ejrnaes, M. J. Rooks, R. Cristiano, S. Pagano, A. Frydman, and D. E. Prober, "Reset dynamics and latching in niobium superconducting nanowire single-photon detectors", Journal of Applied Physics **108**, 084507 (2010).

[59] K. L. Nicolich, C. Cahall, N. T. Islam, G. P. Lafyatis, J. Kim, A. J. Miller, and D. J. Gauthier, "Universal Model for the Turn-On Dynamics of Superconducting Nanowire Single-Photon Detectors", Physical Review Applied **12**, 034020 (2019).

[60] K. K. Berggren, Q.-Y. Zhao, N. Abebe, M. Chen, P. Ravindran, A. McCaughan, and J. C. Bardin, "A superconducting nanowire can be modeled by using SPICE", Superconductor Science and Technology **31**, 055010 (2018).

[61] A. N. McCaughan, D. M. Oh, and S. W. Nam, "A Stochastic SPICE Model for Superconducting Nanowire Single Photon Detectors and Other Nanowire Devices", IEEE Transactions on Applied Superconductivity **29**, 1 (2019).

[62] A. V. Gurevich and R. G. Mints, "Self-heating in normal metals and superconductors", Reviews of Modern Physics **59**, 941 (1987).

[63] Q.-Y. Zhao, D. Zhu, N. Calandri, A. E. Dane, A. N. McCaughan, F. Bellei, H.-Z. Wang, D. F. Santavicca, and K. K. Berggren, "Single-photon imager based on a superconducting nanowire delay line", Nature Photonics **11**, 247 (2017).

[64] Q.-Y. Zhao, D. F. Santavicca, D. Zhu, B. Noble, and K. K. Berggren, "A distributed electrical model for superconducting nanowire single photon detectors", Applied Physics Letters **113**, 082601 (2018).

[65] D. Zhu, M. Colangelo, C. Chen, B. A. Korzh, F. N. C. Wong, M. D. Shaw, and K. K. Berggren, "Resolving Photon Numbers Using a Superconducting Nanowire with Impedance-Matching Taper", Nano Letters **20**, 3858 (2020).

[66] A. Semenov, B. Günther, U. Böttger, H.-W. Hübers, H. Bartolf, A. Engel, A. Schilling, K. Ilin, M. Siegel, R. Schneider, D. Gerthsen, and N. A. Gippius, "Optical and transport properties of ultrathin NbN films and nanostructures", Physical Review B **80**, 054510 (2009).

[67] K. M. Rosfjord, J. K. W. Yang, E. A. Dauler, A. J. Kerman, V. Anant, B. M. Voronov, G. N. Gol'tsman, and K. K. Berggren, "Nanowire Single-photon detector with an integrated optical cavity and anti-reflection coating", Optics Express **14**, 527 (2006).

[68] S. Miki, M. Takeda, M. Fujiwara, M. Sasaki, and Z. Wang, "Compactly packaged superconducting nanowire single-photon detector with an optical cavity for multichannel system", Optics Express **17**, 23557 (2009).

[69] F. Marsili, V. B. Verma, J. A. Stern, S. Harrington, A. E. Lita, T. Gerrits, I. Vayshenker, B. Baek, M. D. Shaw, R. P. Mirin, and S. W. Nam, "Detecting single infrared photons with 93% system efficiency", Nature Photonics **7**, 210 (2013).

[70] W. Zhang, L. You, H. Li, J. Huang, C. Lv, L. Zhang, X. Liu, J. Wu, Z. Wang, and X. Xie, "NbN superconducting nanowire single photon detector with efficiency over 90% at 1550 nm wavelength operational at compact cryocooler temperature", Science China Physics, Mechanics & Astronomy **60**, 120314 (2017).

[71] W. H. P. Pernice, C. Schuck, O. Minaeva, M. Li, G. N. Goltsman, A. V. Sergienko, and H. X. Tang, "High-speed and high-efficiency travelling wave single-photon detectors embedded in nanophotonic circuits", Nature Communications **3**, 1325 (2012).

[72] A. G. Kozorezov, C. Lambert, F. Marsili, M. J. Stevens, V. B. Verma, J. P. Allmaras, M. D. Shaw, R. P. Mirin, and S. W. Nam, "Fano fluctuations in superconducting-nanowire single-photon detectors", Physical Review B **96**, 054507 (2017).

[73] A. D. Semenov, M. Sidorova, M. A. Skvortsov, A. Kuzmin, K. Ilin, and M. Siegel, "Local thermal fluctuations in current-carrying superconducting nanowires", Physical Review B **102**, 184508 (2020).

[74] D. Y. Vodolazov and F. M. Peeters, "Rearrangement of the vortex lattice due to instabilities of vortex flow", Physical Review B **76**, 014521 (2007).

[75] C. Carbillet, V. Cherkez, M. A. Skvortsov, M. V. Feigel'man, F. Debontridder, L. B. Ioffe, V. S. Stolyarov, K. Ilin, M. Siegel, D. Roditchev, T. Cren, and C. Brun, "Spectroscopic evidence for strong correlations between local superconducting gap and local Altshuler-Aronov density of states suppression in ultrathin NbN films", Physical Review B **102**, 024504 (2020).

[76] A. J. Kerman, E. A. Dauler, J. K. W. Yang, K. M. Rosfjord, V. Anant, K. K. Berggren, G. N. Gol'tsman, and B. M. Voronov, "Constriction-limited detection efficiency of superconducting nanowire single-photon detectors", Applied Physics Letters **90**, 101110 (2007).

[77] S.-Z. Lin, O. Ayala-Valenzuela, R. D. McDonald, L. N. Bulaevskii, T. G. Holesinger, F. Ronning, N. R. Weisse-Bernstein, T. L. Williamson, A. H. Mueller, M. A. Hoffbauer, M. W. Rabin, and M. J. Graf, "Characterization of the thin-film NbN superconductor for single-photon detection by transport measurements", Physical Review B **87**, 184507 (2013).

[78] A. Engel, K. Inderbitzin, A. Schilling, R. Lusche, A. Semenov, H.-W. Hübers, D. Henrich, M. Hofherr, K. Il'in, and M. Siegel, "Temperature-Dependence of Detection Efficiency in NbN and TaN SNSPD", IEEE Transactions on Applied Superconductivity **23**, 2300505 (2013).

[79] M. Hofherr, D. Rall, K. Ilin, M. Siegel, A. Semenov, H.-W. Hübers, and N. A. Gippius, "Intrinsic detection efficiency of superconducting nanowire single-photon detectors with different thicknesses", Journal of Applied Physics **108**, 014507 (2010).

[80] R. Lusche, A. Semenov, K. Ilin, M. Siegel, Y. Korneeva, A. Trifonov, A. Korneev, G. Goltsman, D. Vodolazov, and H.-W. Hübers, "Effect of the wire width on the intrinsic detection efficiency of superconducting-nanowire single-photon detectors", Journal of Applied Physics **116**, 043906 (2014).

[81] S. Frasca, B. Korzh, M. Colangelo, D. Zhu, A. E. Lita, J. P. Allmaras, E. E. Wollman, V. B. Verma, A. E. Dane, E. Ramirez, A. D. Beyer, S. W. Nam, A. G. Kozorezov, M. D. Shaw, and K. K. Berggren, "Determining the depairing current in superconducting nanowire single-photon detectors", Physical Review B **100**, 054520 (2019).

[82] J. R. Clem and K. K. Berggren, "Geometry-dependent critical currents in superconducting nanocircuits", Physical Review B **84**, 174510 (2011).

[83] D. Henrich, P. Reichensperger, M. Hofherr, J. M. Meckbach, K. Il'in, M. Siegel, A. Semenov, A. Zotova, and D. Y. Vodolazov, "Geometry-induced reduction of the critical current in superconducting nanowires", Physical Review B **86**, 144504 (2012).

[84] H. L. Hortensius, E. F. C. Driessen, T. M. Klapwijk, K. K. Berggren, and J. R. Clem, "Critical-current reduction in thin superconducting wires due to current crowding", Applied Physics Letters **100**, 182602 (2012).

[85] I. Charaev, A. Semenov, S. Doerner, G. Gomard, K. Ilin, and M. Siegel, "Current dependence of the hot-spot response spectrum of superconducting single-photon detectors with different layouts", Superconductor Science and Technology **30**, 025016 (2017).

[86] R. Baghdadi, E. Schmidt, S. Jahani, I. Charaev, M. G. W. Müller, M. Colangelo, D. Zhu, K. Ilin, A. D. Semenov, Z. Jacob, M. Siegel, and K. K. Berggren, “Enhancing the performance of superconducting nanowire-based detectors with high-filling factor by using variable thickness”, Superconductor Science and Technology **34**, 035010 (2021).

[87] I. Charaev, T. Silbernagel, B. Bachowsky, A. Kuzmin, S. Doerner, K. Ilin, A. Semenov, D. Roditchev, D. Y. Vodolazov, and M. Siegel, “Enhancement of superconductivity in NbN nanowires by negative electron-beam lithography with positive resist”, Journal of Applied Physics **122**, 083901 (2017).

[88] I. Esmaeil Zadeh, J. Chang, J. W. N. Los, S. Gyger, A. W. Elshaari, S. Steinhauer, S. N. Dorenbos, and V. Zwiller, “Superconducting nanowire single-photon detectors: A perspective on evolution, state-of-the-art, future developments, and applications”, Applied Physics Letters **118**, 190502 (2021).

[89] I. Esmaeil Zadeh, J. W. N. Los, R. B. M. Gourgues, V. Steinmetz, G. Bulgarini, S. M. Dobrovolskiy, V. Zwiller, and S. N. Dorenbos, “Single-photon detectors combining high efficiency, high detection rates, and ultra-high timing resolution”, APL Photonics **2**, 111301 (2017).

[90] W. Zhang, Q. Jia, L. You, X. Ou, H. Huang, L. Zhang, H. Li, Z. Wang, and X. Xie, “Saturating Intrinsic Detection Efficiency of Superconducting Nanowire Single-Photon Detectors via Defect Engineering”, Physical Review Applied **12**, 044040 (2019).

[91] J. Chang, J. W. N. Los, J. O. Tenorio-Pearl, N. Noordzij, R. Gourgues, A. Guardiani, J. R. Zichi, S. F. Pereira, H. P. Urbach, V. Zwiller, S. N. Dorenbos, and I. Esmaeil Zadeh, “Detecting telecom single photons with $(99.5^{+0.5}_{-2.07})\%$ system detection efficiency and high time resolution”, APL Photonics **6**, 036114 (2021).

[92] P. Hu, H. Li, L. You, H. Wang, Y. Xiao, J. Huang, X. Yang, W. Zhang, Z. Wang, and X. Xie, “Detecting single infrared photons toward optimal system detection efficiency”, Optics Express **28**, 36884 (2020).

[93] C. M. Natarajan, M. G. Tanner, and R. H. Hadfield, “Superconducting nanowire single-photon detectors: physics and applications”, Superconductor Science and Technology **25**, 063001 (2012).

[94] A. Korneev, Y. Korneeva, I. Florya, B. Voronov, and G. Goltsman, “NbN nanowire superconducting single-photon detector for mid-infrared”, Physics Procedia **36**, 72 (2012).

[95] H. Bartolf, A. Engel, A. Schilling, K. Il’in, M. Siegel, H.-W. Hübers, and A. Semenov, “Current-assisted thermally activated flux liberation in ultrathin nanopatterned NbN superconducting meander structures”, Physical Review B **81**, 024502 (2010).

[96] A. Engel, A. Schilling, K. Il'in, and M. Siegel, "Dependence of count rate on magnetic field in superconducting thin-film TaN single-photon detectors", Physical Review B **86**, 140506 (2012).

[97] A. Semenov, I. Charaev, R. Lusche, K. Ilin, M. Siegel, H.-W. Hübers, N. Bralovi ć, K. Dopf, and D. Y. Vodolazov, "Asymmetry in the effect of magnetic field on photon detection and dark counts in bended nanostrips", Physical Review B **92**, 174518 (2015).

[98] M. Hofherr, D. Rall, K. Il'in, A. Semenov, H.-W. Hübers, and M. Siegel, "Dark Count Suppression in Superconducting Nanowire Single Photon Detectors", Journal of Low Temperature Physics **167**, 822 (2012).

[99] L. You, "Superconducting nanowire single-photon detectors for quantum information", Nanophotonics **9**, 2673 (2020).

[100] A. Divochiy, F. Marsili, D. Bitauld, A. Gaggero, R. Leoni, F. Mattioli, A. Korneev, V. Seleznev, N. Kaurova, O. Minaeva, G. Gol'tsman, K. G. Lagoudakis, M. Benkhaoul, F. Lévy, and A. Fiore, "Superconducting nanowire photon-number-resolving detector at telecommunication wavelengths", Nature Photonics **2**, 302 (2008).

[101] A. Vetter, S. Ferrari, P. Rath, R. Alaee, O. Kahl, V. Kovalyuk, S. Diewald, G. N. Goltsman, A. Korneev, C. Rockstuhl, and W. H. P. Pernice, "Cavity-Enhanced and Ultrafast Superconducting Single-Photon Detectors", Nano Letters **16**, 7085 (2016).

[102] M. Tarkhov, J. Claudon, J. P. Poizat, A. Korneev, A. Divochiy, O. Minaeva, V. Seleznev, N. Kaurova, B. Voronov, A. V. Semenov, and G. Gol'tsman, "Ultrafast reset time of superconducting single photon detectors", Applied Physics Letters **92**, 241112 (2008).

[103] M. Ejrnaes, A. Casaburi, O. Quaranta, S. Marchetti, A. Gaggero, F. Mattioli, R. Leoni, S. Pagano, and R. Cristiano, "Characterization of parallel superconducting nanowire single photon detectors", Superconductor Science and Technology **22**, 055006 (2009).

[104] M. Perrenoud, M. Caloz, E. Amri, C. Autebert, C. Schönenberger, H. Zbinden, and F. Bussières, "Operation of parallel SNSPDs at high detection rates", Superconductor Science and Technology **34**, 024002 (2021).

[105] D. Vodolazov, N. Manova, Y. Korneeva, and A. Korneev, "Timing Jitter in NbN Superconducting Microstrip Single-Photon Detector", Physical Review Applied **14**, 044041 (2020).

[106] W. Zhang, J. Huang, C. Zhang, L. You, C. Lv, L. Zhang, H. Li, Z. Wang, and X. Xie, "A 16-Pixel Interleaved Superconducting Nanowire Single-Photon Detector Array With A Maximum Count Rate Exceeding 1.5 GHz", IEEE Transactions on Applied Superconductivity **29**, 1 (2019).

[107] A. J. Kerman, D. Rosenberg, R. J. Molnar, and E. A. Dauler, "Readout of superconducting nanowire single-photon detectors at high count rates", Journal of Applied Physics **113**, 144511 (2013).

[108] S. Ferrari, V. Kovalyuk, A. Vetter, C. Lee, C. Rockstuhl, A. Semenov, G. Gol'tsman, and W. Pernice, "Analysis of the detection response of waveguide-integrated superconducting nanowire single-photon detectors at high count rate", Applied Physics Letters **115**, 101104 (2019).

[109] A. Kuzmin, S. Doerner, M. Sidorova, S. Wuensch, K. Ilin, M. Siegel, and A. Semenov, "Geometrical Jitter and Bolometric Regime in Photon Detection by Straight Superconducting Nanowire", IEEE Transactions on Applied Superconductivity **29**, 1 (2019).

[110] M. Sidorova, A. Semenov, H.-W. Hübers, I. Charaev, A. Kuzmin, S. Doerner, and M. Siegel, "Physical mechanisms of timing jitter in photon detection by current-carrying superconducting nanowires", Physical Review B **96**, 184504 (2017).

[111] L. You, X. Yang, Y. He, W. Zhang, D. Liu, W. Zhang, L. Zhang, L. Zhang, X. Liu, S. Chen, Z. Wang, and X. Xie, "Jitter analysis of a superconducting nanowire single photon detector", AIP Advances **3**, 072135 (2013).

[112] N. Calandri, Q.-Y. Zhao, D. Zhu, A. Dane, and K. K. Berggren, "Superconducting nanowire detector jitter limited by detector geometry", Applied Physics Letters **109**, 152601 (2016).

[113] M. Sidorova, A. Semenov, H.-W. Hübers, A. Kuzmin, S. Doerner, K. Ilin, M. Siegel, I. Charaev, and D. Vodolazov, "Timing jitter in photon detection by straight superconducting nanowires: Effect of magnetic field and photon flux", Physical Review B **98**, 134504 (2018).

[114] J. Allmaras, A. Kozorezov, B. Korzh, K. Berggren, and M. Shaw, "Intrinsic Timing Jitter and Latency in Superconducting Nanowire Single-photon Detectors", Physical Review Applied **11**, 034062 (2019).

[115] D. Y. Vodolazov, "Minimal Timing Jitter in Superconducting Nanowire Single-Photon Detectors", Physical Review Applied **11**, 014016 (2019).

[116] J. Toussaint, S. Dochow, I. Latka, A. Lukic, T. May, H.-G. Meyer, K. Il'in, M. Siegel, and J. Popp, "Proof of concept of fiber dispersed Raman spectroscopy using superconducting nanowire single-photon detectors", Optics Express **23**, 5078 (2015).

[117] R. Cheng, C.-L. Zou, X. Guo, S. Wang, X. Han, and H. X. Tang, "Broadband on-chip single-photon spectrometer", Nature Communications **10**, 4104 (2019).

[118] C. Cahall, K. L. Nicolich, N. T. Islam, G. P. Lafyatis, A. J. Miller, D. J. Gauthier, and J. Kim, "Multi-photon detection using a conventional superconducting nanowire single-photon detector", Optica **4**, 1534 (2017).

[119] V. Anant, A. J. Kerman, E. A. Dauler, J. K. W. Yang, K. M. Rosfjord, and K. K. Berggren, "Optical properties of superconducting nanowire single-photon detectors", Optics Express **16**, 10750 (2008).

[120] Q. Guo, H. Li, L. You, W. Zhang, L. Zhang, Z. Wang, X. Xie, and M. Qi, "Single photon detector with high polarization sensitivity", Scientific Reports **5**, 9616 (2015).

[121] R. Xu, Y. Li, F. Zheng, G. Zhu, L. Kang, L. Zhang, X. Jia, X. Tu, Q. Zhao, B. Jin, W. Xu, J. Chen, and P. Wu, "Demonstration of a superconducting nanowire single photon detector with an ultrahigh polarization extinction ratio over 400", Optics Express **26**, 3947 (2018).

[122] S. N. Dorenbos, E. M. Reiger, N. Akopian, U. Perinetti, V. Zwiller, T. Zijlstra, and T. M. Klapwijk, "Superconducting single photon detectors with minimized polarization dependence", Applied Physics Letters **93**, 161102 (2008).

[123] V. B. Verma, F. Marsili, S. Harrington, A. E. Lita, R. P. Mirin, and S. W. Nam, "A three-dimensional, polarization-insensitive superconducting nanowire avalanche photodetector", Applied Physics Letters **101**, 251114 (2012).

[124] X. Chi, K. Zou, C. Gu, J. Zichi, Y. Cheng, N. Hu, X. Lan, S. Chen, Z. Lin, V. Zwiller, and X. Hu, "Fractal superconducting nanowire single-photon detectors with reduced polarization sensitivity", Optics Letters **43**, 5017 (2018).

[125] F. Zheng, R. Xu, G. Zhu, B. Jin, L. Kang, W. Xu, J. Chen, and P. Wu, "Design of a polarization-insensitive superconducting nanowire single photon detector with high detection efficiency", Scientific Reports **6**, 22710 (2016).

[126] S. Miki, M. Takeda, M. Fujiwara, M. Sasaki, A. Otomo, and Z. Wang, "Superconducting NbTiN Nanowire Single Photon Detectors with Low Kinetic Inductance", Applied Physics Express **2**, 075002 (2009).

[127] S. Miki, T. Yamashita, H. Terai, and Z. Wang, "High performance fiber-coupled NbTiN superconducting nanowire single photon detectors with Gifford-McMahon cryocooler", Optics Express **21**, 10208 (2013).

[128] C. Schuck, W. H. P. Pernice, and H. X. Tang, "Waveguide integrated low noise NbTiN nanowire single-photon detectors with milli-Hz dark count rate", Scientific Reports **3**, 1893 (2013).

[129] X. Yang, L. You, L. Zhang, C. Lv, H. Li, X. Liu, H. Zhou, and Z. Wang, "Comparison of Superconducting Nanowire Single-Photon Detectors Made of NbTiN and NbN Thin Films", IEEE Transactions on Applied Superconductivity **28**, 1 (2018).

[130] K. Il'in, M. Hofherr, D. Rall, M. Siegel, A. Semenov, A. Engel, K. Inderbitzin, A. Aeschbacher, and A. Schilling, "Ultra-thin TaN Films for Superconducting Nanowire Single-Photon Detectors", Journal of Low Temperature Physics **167**, 809 (2012).

[131] K. S. Il'in, D. Rall, M. Siegel, and A. Semenov, "Critical current density in thin superconducting TaN film structures", Physica C: Superconductivity **479**, 176 (2012).

[132] A. Engel, A. Aeschbacher, K. Inderbitzin, A. Schilling, K. Il'in, M. Hofherr, M. Siegel, A. Semenov, and H.-W. Hübers, "Tantalum nitride superconducting single-photon detectors with low cut-off energy", Applied Physics Letters **100**, 062601 (2012).

[133] D. Morozov, A. Banerjee, K. Erotokritou, G. Taylor, N. R. Gemmell, D. Hemakumara, I. Thayne, and R. H. Hadfield, "Superconducting nanowire materials for mid infrared single photon detection (Conference Presentation)", Proceedings of SPIE **10659**, 106590G (2018).

[134] A. Engel, A. Semenov, H. W. Hübers, K. Il'in, and M. Siegel, "Superconducting single-photon detector for the visible and infrared spectral range", Journal of Modern Optics **51**, 1459 (2004).

[135] A. J. Annunziata, D. F. Santavicca, J. D. Chudow, L. Frunzio, M. J. Rooks, A. Frydman, and D. E. Prober, "Niobium Superconducting Nanowire Single-Photon Detectors", IEEE Transactions on Applied Superconductivity **19**, 327 (2009).

[136] B. Baek, A. E. Lita, V. Verma, and S. W. Nam, "Superconducting a-W_xSi_{1-x} nanowire single-photon detector with saturated internal quantum efficiency from visible to 1850 nm", Applied Physics Letters **98**, 251105 (2011).

[137] J. Chiles, S. M. Buckley, A. Lita, V. B. Verma, J. Allmaras, B. Korzh, M. D. Shaw, J. M. Shainline, R. P. Mirin, and S. W. Nam, "Superconducting microwire detectors based on WSi with single-photon sensitivity in the near-infrared", Applied Physics Letters **116**, 242602 (2020).

[138] Y. P. Korneeva, M. Y. Mikhailov, Y. P. Pershin, N. N. Manova, A. V. Divochiy, Y. B. Vakhtomin, A. A. Korneev, K. V. Smirnov, A. G. Sivakov, A. Y. Devizenko, and G. N. Goltsman, "Superconducting single-photon detector made of MoSi film", Superconductor Science and Technology **27**, 095012 (2014).

[139] V. B. Verma, B. Korzh, F. Bussières, R. D. Horansky, S. D. Dyer, A. E. Lita, I. Vayshenker, F. Marsili, M. D. Shaw, H. Zbinden, R. P. Mirin, and S. W. Nam, "High-efficiency superconducting nanowire single-photon detectors fabricated from MoSi thin-films", Optics Express **23**, 33792 (2015).

[140] A. E. Lita, V. B. Verma, J. Chiles, R. P. Mirin, and S. W. Nam, "Mo_xSi_{1-x} a versatile material for nanowire to microwire single-photon detectors from UV to near IR", Superconductor Science and Technology **34**, 054001 (2021).

[141] I. Charaev, Y. Morimoto, A. Dane, A. Agarwal, M. Colangelo, and K. K. Berggren, "Large-area microwire MoSi single-photon detectors at 1550 nm wavelength", Applied Physics Letters **116**, 242603 (2020).

[142] S. N. Dorenbos, P. Forn-Diaz, T. Fuse, A. H. Verbruggen, T. Zijlstra, T. M. Klapwijk, and V. Zwiller, "Low gap superconducting single photon detectors for infrared sensitivity", Applied Physics Letters **98**, 251102 (2011).

[143] V. B. Verma, A. E. Lita, M. R. Vissers, F. Marsili, D. P. Pappas, R. P. Mirin, and S. W. Nam, "Superconducting nanowire single photon detectors fabricated from an amorphous $Mo_{0.75}Ge_{0.25}$ thin film", Applied Physics Letters **105**, 022602 (2014).

[144] X. Zhang, A. Engel, Q. Wang, A. Schilling, A. Semenov, M. Sidorova, H.-W. Hübers, I. Charaev, K. Ilin, and M. Siegel, "Characteristics of superconducting tungsten silicide W_xSi_{1-x} for single photon detection", Physical Review B **94**, 174509 (2016).

[145] H. Shibata, H. Takesue, T. Honjo, T. Akazaki, and Y. Tokura, "Single-photon detection using magnesium diboride superconducting nanowires", Applied Physics Letters **97**, 212504 (2010).

[146] H. Shibata, T. Akazaki, and Y. Tokura, "Fabrication of MgB_2 Nanowire Single-Photon Detector with Meander Structure", Applied Physics Express **6**, 023101 (2013).

[147] S. Cherednichenko, N. Acharya, E. Novoselov, and V. Drakinskiy, "Low kinetic inductance superconducting MgB_2 nanowires with a 130 ps relaxation time for single-photon detection applications", Superconductor Science and Technology **34**, 044001 (2021).

[148] R. Arpaia, D. Golubev, R. Baghdadi, R. Ciancio, G. Dražić, P. Orgiani, D. Montemurro, T. Bauch, and F. Lombardi, "Transport properties of ultrathin $YBa_2Cu_3O_{7-\delta}$ nanowires: A route to single-photon detection", Physical Review B **96**, 064525 (2017).

[149] M. Ejrnaes, L. Parlato, R. Arpaia, T. Bauch, F. Lombardi, R. Cristiano, F. Tafuri, and G. P. Pepe, "Observation of dark pulses in 10 nm thick YBCO nanostrips presenting hysteretic current voltage characteristics", Superconductor Science and Technology **30**, 12LT02 (2017).

[150] P. Amari, C. Feuillet-Palma, A. Jouan, F. Couëdo, N. Bourlet, E. Géron, M. Malnou, L. Méchin, A. Sharafiev, J. Lesueur, and N. Bergeal, "High-temperature superconducting nano-meanders made by ion irradiation", Superconductor Science and Technology **31**, 015019 (2017).

[151] K. Ilin, D. Henrich, Y. Luck, Y. Liang, M. Siegel, and D. Y. Vodolazov, "Critical current of Nb, NbN, and TaN thin-film bridges with and without geometrical nonuniformities in a magnetic field", Physical Review B **89**, 184511 (2014).

[152] D. Hazra, N. Tsavdaris, A. Mukhtarova, M. Jacquemin, F. Blanchet, R. Albert, S. Jebari, A. Grimm, A. Konar, E. Blanquet, F. Mercier, C. Chapelier, and M. Hofheinz, "Superconducting properties of NbTiN thin films deposited by high-temperature chemical vapor deposition", Physical Review B **97**, 144518 (2018).

[153] M. V. Sidorova, A. G. Kozorezov, A. V. Semenov, Y. P. Korneeva, M. Y. Mikhailov, A. Y. Devizenko, A. A. Korneev, G. M. Chulkova, and G. N. Goltsman, "Nonbolometric bottleneck in electron-phonon relaxation in ultrathin WSi films", Physical Review B **97**, 184512 (2018).

[154] I. Holzman and Y. Ivry, "Superconducting Nanowires for Single-Photon Detection: Progress, Challenges, and Opportunities", Advanced Quantum Technologies **2**, 1800058 (2019).

[155] H. Shibata, "Review of Superconducting Nanostrip Photon Detectors using Various Superconductors", IEICE Transactions on Electronics **E104.C**, 429 (2021).

[156] D. Henrich, S. Dörner, M. Hofherr, K. Il'in, A. Semenov, E. Heintze, M. Scheffler, M. Dressel, and M. Siegel, "Broadening of hot-spot response spectrum of superconducting NbN nanowire single-photon detector with reduced nitrogen content", Journal of Applied Physics **112**, 074511 (2012).

[157] D. Henrich, "Influence of Material and Geometry on the Performance of Superconducting Nanowire Single-Photon Detectors", PhD thesis (Karlsruher Institut für Technologie, 2013).

[158] A. E. Dane, A. N. McCaughan, D. Zhu, Q. Zhao, C.-S. Kim, N. Calandri, A. Agarwal, F. Bellei, and K. K. Berggren, "Bias sputtered NbN and superconducting nanowire devices", Applied Physics Letters **111**, 122601 (2017).

[159] R. W. Johnson, A. Hultqvist, and S. F. Bent, "A brief review of atomic layer deposition: from fundamentals to applications", Materials Today **17**, 236 (2014).

[160] H. B. Profijt, S. E. Potts, M. C. M. van de Sanden, and W. M. M. Kessels, "Plasma-Assisted Atomic Layer Deposition: Basics, Opportunities, and Challenges", Journal of Vacuum Science & Technology A **29**, 050801 (2011).

[161] H. C. M. Knoops, T. Faraz, K. Arts, and W. M. M. Kessels, "Status and prospects of plasma-assisted atomic layer deposition", Journal of Vacuum Science & Technology A **37**, 030902 (2019).

[162] L. Hiltunen, M. Leskelä, M. Mäkelä, L. Niinistö, E. Nykänen, and P. Soininen, "Nitrides of titanium, niobium, tantalum and molybdenum grown as thin films by the Atomic Layer Epitaxy method", Thin Solid Films **166**, 149 (1988).

[163] K.-E. Elers, M. Ritala, M. Leskelä, and E. Rauhala, "$NbCl_5$ as a precursor in atomic layer epitaxy", Applied Surface Science **82-83**, 468 (1994).

[164] M. Ritala, T. Asikainen, M. Leskelä, J. Jokinen, R. Lappalainen, M. Utriainen, L. Niinistö, and E. Ristolainen, "Effects of intermediate zinc pulses on properties of TiN and NbN films deposited by atomic layer epitaxy", Applied Surface Science **120**, 199 (1997).

[165] M. Juppo, M. Ritala, and M. Leskelä, "Use of 1,1-Dimethylhydrazine in the Atomic Layer Deposition of Transition Metal Nitride Thin Films", Journal of The Electrochemical Society **147**, 3377 (2000).

[166] P. Alén, M. Ritala, K. Arstila, J. Keinonen, and M. Leskelä, "The growth and diffusion barrier properties of atomic layer deposited NbN_x thin films", Thin Solid Films **491**, 235 (2005).

[167] N. Van Hoornick, H. De Witte, T. Witters, C. Zhao, T. Conard, H. Huotari, J. Swerts, T. Schram, J. W. Maes, S. De Gendt, and M. Heyns, "Evaluation of Atomic Layer Deposited NbN and NbSiN as Metal Gate Materials", Journal of The Electrochemical Society **153**, G437 (2006).

[168] J. Hinz, A. J. Bauer, and L. Frey, "Analysis of NbN thin film deposition by plasma-enhanced ALD for gate electrode application", Semiconductor Science and Technology **25**, 075009 (2010).

[169] M. Ziegler, L. Fritzsch, J. Day, S. Linzen, S. Anders, J. Toussaint, and H.-G. Meyer, "Superconducting niobium nitride thin films deposited by metal organic plasma-enhanced atomic layer deposition", Superconductor Science and Technology **26**, 025008 (2013).

[170] M. J. Sowa, Y. Yemane, J. Zhang, J. C. Palmstrom, L. Ju, N. C. Strandwitz, F. B. Prinz, and J. Provine, "Plasma-enhanced atomic layer deposition of superconducting niobium nitride", Journal of Vacuum Science & Technology A **35**, 01B143 (2017).

[171] M. Ziegler, S. Linzen, S. Goerke, U. Brückner, J. Plentz, J. Dellith, A. Himmerlich, M. Himmerlich, U. Hübner, S. Krischok, and H.-G. Meyer, "Effects of Plasma Parameter on Morphological and Electrical Properties of Superconducting Nb-N Deposited by MO-PEALD", IEEE Transactions on Applied Superconductivity **27**, 1 (2017).

[172] S. Linzen, M. Ziegler, O. V. Astafiev, M. Schmelz, U. Hübner, M. Diegel, E. Il'ichev, and H.-G. Meyer, "Structural and electrical properties of ultrathin niobium nitride films grown by atomic layer deposition", Superconductor Science and Technology **30**, 035010 (2017).

[173] V. B. Verma, A. E. Lita, B. Korzh, E. Wollman, M. D. Shaw, R. P. Mirin, and S. W. Nam, "Towards single-photon spectroscopy in the mid-infrared using superconducting nanowire single-photon detectors", Proceedings of SPIE **10978**, 109780N (2019).

[174] V. B. Verma, R. Horansky, F. Marsili, J. A. Stern, M. D. Shaw, A. E. Lita, R. P. Mirin, and S. W. Nam, "A four-pixel single-photon pulse-position array fabricated from WSi superconducting nanowire single-photon detectors", Applied Physics Letters **104**, 051115 (2014).

[175] M. S. Allman, V. B. Verma, M. Stevens, T. Gerrits, R. D. Horansky, A. E. Lita, F. Marsili, A. Beyer, M. D. Shaw, D. Kumor, R. Mirin, and S. W. Nam, "A near-infrared 64-pixel superconducting nanowire single photon detector array with integrated multiplexed readout", Applied Physics Letters **106**, 192601 (2015).

[176] J. P. Allmaras, E. E. Wollman, A. D. Beyer, R. M. Briggs, B. A. Korzh, B. Bumble, and M. D. Shaw, "Demonstration of a Thermally Coupled Row-Column SNSPD Imaging Array", Nano Letters **20**, 2163 (2020).

[177] M. Hofherr, M. Arndt, K. Il'in, D. Henrich, M. Siegel, J. Toussaint, T. May, and H.-G. Meyer, "Time-Tagged Multiplexing of Serially Biased Superconducting Nanowire Single-Photon Detectors", IEEE Transactions on Applied Superconductivity **23**, 2501205 (2013).

[178] D. Zhu, Q.-Y. Zhao, H. Choi, T.-J. Lu, A. E. Dane, D. Englund, and K. K. Berggren, "A scalable multi-photon coincidence detector based on superconducting nanowires", Nature Nanotechnology **13**, 596 (2018).

[179] A. Gaggero, F. Martini, F. Mattioli, F. Chiarello, R. Cernansky, A. Politi, and R. Leoni, "Amplitude-multiplexed readout of single photon detectors based on superconducting nanowires", Optica **6**, 823 (2019).

[180] H. Terai, S. Miki, T. Yamashita, K. Makise, and Z. Wang, "Demonstration of single-flux-quantum readout operation for superconducting single-photon detectors", Applied Physics Letters **97**, 112510 (2010).

[181] T. Yamashita, S. Miki, H. Terai, K. Makise, and Z. Wang, "Crosstalk-free operation of multielement superconducting nanowire single-photon detector array integrated with single-flux-quantum circuit in a 0.1 W Gifford-McMahon cryocooler", Optics Letters **37**, 2982 (2012).

[182] S. Miyajima, M. Yabuno, S. Miki, T. Yamashita, and H. Terai, “High-time-resolved 64-channel single-flux quantum-based address encoder integrated with a multi-pixel superconducting nanowire single-photon detector”, Optics Express **26**, 29045 (2018).

[183] M. A. Dobbs, M. Lueker, K. A. Aird, A. N. Bender, B. A. Benson, L. E. Bleem, J. E. Carlstrom, C. L. Chang, H.-M. Cho, J. Clarke, T. M. Crawford, A. T. Crites, D. I. Flanigan, T. de Haan, E. M. George, N. W. Halverson, W. L. Holzapfel, J. D. Hrubes, B. R. Johnson, J. Joseph, R. Keisler, J. Kennedy, Z. Kermish, T. M. Lanting, A. T. Lee, E. M. Leitch, D. Luong-Van, J. J. McMahon, J. Mehl, S. S. Meyer, T. E. Montroy, S. Padin, T. Plagge, C. Pryke, P. L. Richards, J. E. Ruhl, K. K. Schaffer, D. Schwan, E. Shirokoff, H. G. Spieler, Z. Staniszewski, A. A. Stark, K. Vanderlinde, J. D. Vieira, C. Vu, B. Westbrook, and R. Williamson, “Frequency multiplexed superconducting quantum interference device readout of large bolometer arrays for cosmic microwave background measurements”, Review of Scientific Instruments **83**, 073113 (2012).

[184] P. K. Day, H. G. LeDuc, B. A. Mazin, A. Vayonakis, and J. Zmuidzinas, “A broadband superconducting detector suitable for use in large arrays”, Nature **425**, 817 (2003).

[185] S. McHugh, B. A. Mazin, B. Serfass, S. Meeker, K. O’Brien, R. Duan, R. Raffanti, and D. Werthimer, “A readout for large arrays of microwave kinetic inductance detectors”, Review of Scientific Instruments **83**, 044702 (2012).

[186] S. Doerner, A. Kuzmin, S. Wuensch, I. Charaev, F. Boes, T. Zwick, and M. Siegel, “Frequency-multiplexed bias and readout of a 16-pixel superconducting nanowire single-photon detector array”, Applied Physics Letters **111**, 032603 (2017).

[187] A. K. Sinclair, E. Schroeder, D. Zhu, M. Colangelo, J. Glasby, P. D. Mauskopf, H. Mani, and K. K. Berggren, “Demonstration of Microwave Multiplexed Readout of DC-Biased Superconducting Nanowire Detectors”, IEEE Transactions on Applied Superconductivity **29**, 1 (2019).

[188] S. Doerner, A. Kuzmin, S. Wuensch, I. Charaev, and M. Siegel, “Operation of Multipixel Radio-Frequency Superconducting Nanowire Single-Photon Detector Arrays”, IEEE Transactions on Applied Superconductivity **27**, 1 (2017).

[189] S. Doerner, A. Kuzmin, S. Wuensch, and M. Siegel, “Comparison of SNSPDs Biased With Microwave and Direct Currents”, IEEE Transactions on Applied Superconductivity **29**, 1 (2019).

[190] S. Doerner, “Multifrequenzausleseverfahren von supraleitenden Einzelphotonen-Detektoren”, PhD thesis (Karlsruher Institut für Technologie, 2019).

[191] D. Zhu, M. Colangelo, B. A. Korzh, Q.-Y. Zhao, S. Frasca, A. E. Dane, A. E. Velasco, A. D. Beyer, J. P. Allmaras, E. Ramirez, W. J. Strickland, D. F. Santavicca, M. D. Shaw, and K. K. Berggren, "Superconducting nanowire single-photon detector with integrated impedance-matching taper", Applied Physics Letters **114**, 042601 (2019).

[192] R. Duan, S. McHugh, B. Serfass, B. A. Mazin, A. Merrill, S. R. Golwala, T. P. Downes, N. G. Czakon, P. K. Day, J. Gao, J. Glenn, M. I. Hollister, H. G. Leduc, P. R. Maloney, O. Noroozian, H. T. Nguyen, J. Sayers, J. A. Schlaerth, S. Siegel, J. E. Vaillancourt, A. Vayonakis, P. R. Wilson, and J. Zmuidzinas, "An open-source readout for MKIDs", Proceedings of SPIE **7741**, 77411V (2010).

[193] P.-I. Dietrich, M. Blaicher, I. Reuter, M. Billah, T. Hoose, A. Hofmann, C. Caer, R. Dangel, B. Offrein, U. Troppenz, M. Moehrle, W. Freude, and C. Koos, "In situ 3D nanoprinting of free-form coupling elements for hybrid photonic integration", Nature Photonics **12**, 241 (2018).

[194] A. J. Miller, A. E. Lita, B. Calkins, I. Vayshenker, S. M. Gruber, and S. W. Nam, "Compact cryogenic self-aligning fiber-to-detector coupling with losses below one percent", Optics Express **19**, 9102 (2011).

[195] Y. Xu, J. Wu, W. Fang, L. You, and L. Tong, "Microfiber coupled superconducting nanowire single-photon detectors", Optics Communications **405**, 48 (2017).

[196] L. You, J. Wu, Y. Xu, X. Hou, W. Fang, H. Li, W. Zhang, L. Zhang, X. Liu, L. Tong, Z. Wang, and X. Xie, "Microfiber-coupled superconducting nanowire single-photon detector for near-infrared wavelengths", Optics Express **25**, 31221 (2017).

[197] X. Hou, N. Yao, L. You, H. Li, Y. Wang, W. Zhang, H. Wang, X. Liu, W. Fang, L. Tong, Z. Wang, and X. Xie, "Ultra-broadband microfiber-coupled superconducting single-photon detector", Optics Express **27**, 25241 (2019).

[198] W.-J. Zhang, G.-Z. Xu, L.-X. You, C.-J. Zhang, H. Huang, X. Ou, X.-Q. Sun, J.-M. Xiong, H. Li, Z. Wang, and X.-M. Xie, "Sixteen-channel fiber array-coupled superconducting single-photon detector array with average system detection efficiency over 60% at telecom wavelength", Optics Letters **46**, 1049 (2021).

[199] W. J. Zhang, X. Y. Yang, H. Li, L. X. You, C. L. Lv, L. Zhang, C. J. Zhang, X. Y. Liu, Z. Wang, and X. M. Xie, "Fiber-coupled superconducting nanowire single-photon detectors integrated with a bandpass filter on the fiber end-face", Superconductor Science and Technology **31**, 035012 (2018).

[200] H. Shibata, T. Hiraki, T. Tsuchizawa, K. Yamada, Y. Tokura, and S. Matsuo, "A waveguide-integrated superconducting nanowire single-photon detector with a spot-size converter on a Si photonics platform", Superconductor Science and Technology **32**, 034001 (2019).

[201] N. Lindenmann, G. Balthasar, D. Hillerkuss, R. Schmogrow, M. Jordan, J. Leuthold, W. Freude, and C. Koos, "Photonic wire bonding: a novel concept for chip-scale interconnects", Optics Express **20**, 17667 (2012).

[202] N. Lindenmann, S. Dottermusch, M. L. Goedecke, T. Hoose, M. R. Billah, T. P. Onanuga, A. Hofmann, W. Freude, and C. Koos, "Connecting Silicon Photonic Circuits to Multicore Fibers by Photonic Wire Bonding", Journal of Lightwave Technology **33**, 755 (2015).

[203] M. Schwartz, E. Schmidt, U. Rengstl, F. Hornung, S. Hepp, S. L. Portalupi, K. llin, M. Jetter, M. Siegel, and P. Michler, "Fully On-Chip Single-Photon Hanbury-Brown and Twiss Experiment on a Monolithic Semiconductor–Superconductor Platform", Nano Letters **18**, 6892 (2018).

[204] M. R. Billah, M. Blaicher, T. Hoose, P.-I. Dietrich, P. Marin-Palomo, N. Lindenmann, A. Nesic, A. Hofmann, U. Troppenz, M. Moehrle, S. Randel, W. Freude, and C. Koos, "Hybrid integration of silicon photonics circuits and InP lasers by photonic wire bonding", Optica **5**, 876 (2018).

[205] M. Blaicher, M. R. Billah, J. Kemal, T. Hoose, P. Marin-Palomo, A. Hofmann, Y. Kutuvantavida, C. Kieninger, P.-I. Dietrich, M. Lauermann, S. Wolf, U. Troppenz, M. Moehrle, F. Merget, S. Skacel, J. Witzens, S. Randel, W. Freude, and C. Koos, "Hybrid multi-chip assembly of optical communication engines by in situ 3D nano-lithography", Light: Science & Applications **9**, 71 (2020).

[206] M. D. Eisaman, J. Fan, A. Migdall, and S. V. Polyakov, "Invited Review Article: Single-photon sources and detectors", Review of Scientific Instruments **82**, 071101 (2011).

[207] H. Knoops, S. Potts, A. Bol, and W. Kessels, "Atomic Layer Deposition", in *Thin Films and Epitaxy*, edited by T. F. Kuech, 2nd edition, Handbook of Crystal Growth (Elsevier, 2015).

[208] N. E. Richey, C. de Paula, and S. F. Bent, "Understanding chemical and physical mechanisms in atomic layer deposition", The Journal of Chemical Physics **152**, 040902 (2020).

[209] E. Knehr, A. Kuzmin, D. Y. Vodolazov, M. Ziegler, S. Doerner, K. Ilin, M. Siegel, R. Stolz, and H. Schmidt, "Nanowire single-photon detectors made of atomic layer-deposited niobium nitride", Superconductor Science and Technology **32**, 125007 (2019).

[210] E. Knehr, "Entwicklung Niobnitrid-basierter Einzelphotonendetektoren mittels plasmagestützter Atomlagenabscheidung", master's thesis (Karlsruhe Institute of Technology, 2018).

[211] Y. V. Fominov and M. V. Feigel'man, "Superconductive properties of thin dirty superconductor–normal-metal bilayers", Physical Review B **63**, 094518 (2001).

[212] K. Ilin, R. Schneider, D. Gerthsen, A. Engel, H. Bartolf, A. Schilling, A. Semenov, H.-W. Huebers, B. Freitag, and M. Siegel, "Ultra-thin NbN films on Si: crystalline and superconducting properties", Journal of Physics: Conference Series **97**, 012045 (2008).

[213] A. Engel, A. Semenov, H.-W. Hübers, K. Il'in, and M. Siegel, "Electric noise and local photon-induced nonequilibrium states in a current-carrying nanostructured superconductor", in *New Frontiers in Superconductivity Research*, edited by B. P. Martins, 1st edition (Nova Science Publishers, 2006).

[214] E. Helfand and N. R. Werthamer, "Temperature and Purity Dependence of the Superconducting Critical Field, H_{c2}. II", Physical Review **147**, 288 (1966).

[215] J. Pearl, "Current distribution in superconducting films carrying quantized fluxoids", Applied Physics Letters **5**, 65 (1964).

[216] K. Il'in, M. Siegel, A. Semenov, A. Engel, and H.-W. Hübers, "Critical current of Nb and NbN thin-film structures: The cross-section dependence", physica status solidi (c) **2**, 1680 (2005).

[217] K. Il'in, D. Rall, M. Siegel, A. Engel, A. Schilling, A. Semenov, and H.-W. Huebers, "Influence of thickness, width and temperature on critical current density of Nb thin film structures", Physica C: Superconductivity **470**, 953 (2010).

[218] M. Kupriyanov and V. Lukichev, "Temperature dependence of pair-breaking current in superconductors", Soviet Journal of Low Temperature Physics **6**, 210 (1980).

[219] E. Knehr, M. Ziegler, S. Linzen, K. Ilin, P. Schanz, J. Plentz, M. Diegel, H. Schmidt, E. Il'ichev, and M. Siegel, "Wafer-level uniformity of atomic-layer-deposited niobium nitride thin films for quantum devices", Journal of Vacuum Science & Technology A **39**, 052401 (2021).

[220] J. A. O'Connor, M. G. Tanner, C. M. Natarajan, G. S. Buller, R. J. Warburton, S. Miki, Z. Wang, S. W. Nam, and R. H. Hadfield, "Spatial dependence of output pulse delay in a niobium nitride nanowire superconducting single-photon detector", Applied Physics Letters **98**, 201116 (2011).

[221] Y. Cheng, C. Gu, and X. Hu, "Inhomogeneity-induced timing jitter of superconducting nanowire single-photon detectors", Applied Physics Letters **111**, 062604 (2017).

[222] B. Sacépé, C. Chapelier, T. I. Baturina, V. M. Vinokur, M. R. Baklanov, and M. Sanquer, "Disorder-Induced Inhomogeneities of the Superconducting State Close to the Superconductor-Insulator Transition", Physical Review Letters **101**, 157006 (2008).

[223] Y. Noat, V. Cherkez, C. Brun, T. Cren, C. Carbillet, F. Debontridder, K. Ilin, M. Siegel, A. Semenov, H.-W. Hübers, and D. Roditchev, "Unconventional superconductivity in ultrathin superconducting NbN films studied by scanning tunneling spectroscopy", Physical Review B **88**, 014503 (2013).

[224] Z. Wang, A. Kawakami, Y. Uzawa, and B. Komiyama, "Superconducting properties and crystal structures of single-crystal niobium nitride thin films deposited at ambient substrate temperature", Journal of Applied Physics **79**, 7837 (1996).

[225] R. Cheng, S. Wang, and H. X. Tang, "Superconducting nanowire single-photon detectors fabricated from atomic-layer-deposited NbN", Applied Physics Letters **115**, 241101 (2019).

[226] J. R. Clem and V. G. Kogan, "Kinetic impedance and depairing in thin and narrow superconducting films", Physical Review B **86**, 174521 (2012).

[227] J. Bardeen, "Critical Fields and Currents in Superconductors", Reviews of Modern Physics **34**, 667 (1962).

[228] K. Il'in, M. Siegel, A. Engel, H. Bartolf, A. Schilling, A. Semenov, and H.-W. Huebers, "Current-Induced Critical State in NbN Thin-Film Structures", Journal of Low Temperature Physics **151**, 585 (2008).

[229] K. K. Likharev, "Superconducting weak links", Reviews of Modern Physics **51**, 101 (1979).

[230] I. Zailer, J. E. F. Frost, V. Chabasseur-Molyneux, C. J. B. Ford, and M. Pepper, "Crosslinked PMMA as a high-resolution negative resist for electron beam lithography and applications for physics of low-dimensional structures", Semiconductor Science and Technology **11**, 1235 (1996).

[231] Y. Korneeva, M. Sidorova, A. Semenov, S. Krasnosvobodtsev, K. Mitsen, A. Korneev, G. Chulkova, and G. Goltsman, "Comparison of Hot-Spot Formation in NbC and NbN Single-Photon Detectors", IEEE Transactions on Applied Superconductivity **26**, 1 (2016).

[232] L. Maingault, M. Tarkhov, I. Florya, A. Semenov, R. Espiau de Lamaëstre, P. Cavalier, G. Gol'tsman, J.-P. Poizat, and J.-C. Villégier, "Spectral dependency of superconducting single photon detectors", Journal of Applied Physics **107**, 116103 (2010).

[233] D. Y. Vodolazov, "Saddle point states in two-dimensional superconducting films biased near the depairing current", Physical Review B **85**, 174507 (2012).

[234] A. Engel, A. Semenov, H.-W. Hübers, K. Ilin, and M. Siegel, "Fluctuations and dark count rates in superconducting NbN single-photon detectors", physica status solidi (c) **2**, 1668 (2005).

[235] J. Kitaygorsky, I. Komissarov, A. Jukna, D. Pan, O. Minaeva, N. Kaurova, A. Divochiy, A. Korneev, M. Tarkhov, B. Voronov, I. Milostnaya, G. Gol'tsman, and R. R. Sobolewski, "Dark Counts in Nanostructured NbN Superconducting Single-Photon Detectors and Bridges", IEEE Transactions on Applied Superconductivity **17**, 275 (2007).

[236] T. Yamashita, S. Miki, K. Makise, W. Qiu, H. Terai, M. Fujiwara, M. Sasaki, and Z. Wang, "Origin of intrinsic dark count in superconducting nanowire single-photon detectors", Applied Physics Letters **99**, 161105 (2011).

[237] J. S. Langer and V. Ambegaokar, "Intrinsic Resistive Transition in Narrow Superconducting Channels", Physical Review **164**, 498 (1967).

[238] D. E. McCumber and B. I. Halperin, "Time Scale of Intrinsic Resistive Fluctuations in Thin Superconducting Wires", Physical Review B **1**, 1054 (1970).

[239] P. M. Marychev and D. Y. Vodolazov, "Threshold fluctuations in a superconducting current-carrying bridge", Superconductor Science and Technology **30**, 075008 (2017).

[240] C. Qiu and T. Qian, "Numerical study of the phase slip in two-dimensional superconducting strips", Physical Review B **77**, 174517 (2008).

[241] E. Knehr, A. Kuzmin, S. Doerner, S. Wuensch, K. Ilin, H. Schmidt, and M. Siegel, "Synchronous single-photon detection with self-resetting, GHz-gated superconducting NbN nanowires", Applied Physics Letters **117**, 132602 (2020).

[242] A. G. Kozorezov, C. Lambert, F. Marsili, M. J. Stevens, V. B. Verma, J. A. Stern, R. Horansky, S. Dyer, S. Duff, D. P. Pappas, A. Lita, M. D. Shaw, R. P. Mirin, and S. W. Nam, "Quasiparticle recombination in hotspots in superconducting current-carrying nanowires", Physical Review B **92**, 064504 (2015).

[243] L. Zhang, S. Zhang, X. Tao, G. Zhu, L. Kang, J. Chen, and P. Wu, "Quasi-Gated Superconducting Nanowire Single-Photon Detector", IEEE Transactions on Applied Superconductivity **27**, 1 (2017).

[244] M. Elezov, R. Ozhegov, G. Goltsman, and V. Makarov, "Countermeasure against bright-light attack on superconducting nanowire single-photon detector in quantum key distribution", Optics Express **27**, 30979 (2019).

[245] A. A. Kuzmin, A. D. Semenov, S. V. Shitov, M. Merker, S. H. Wuensch, A. V. Ustinov, and M. Siegel, "Superconducting noise bolometer with microwave bias and readout for array applications", Applied Physics Letters **111**, 042601 (2017).

[246] J. Zhang, M. A. Itzler, H. Zbinden, and J.-W. Pan, "Advances in InGaAs/InP single-photon detector systems for quantum communication", Light: Science & Applications **4**, e286 (2015).

[247] M. K. Akhlaghi and A. H. Majedi, "Gated mode superconducting nanowire single photon detectors", Optics Express **20**, 1608 (2012).

[248] L. Mandel, "Fluctuations of Photon Beams: The Distribution of the Photo-Electrons", Proceedings of the Physical Society **74**, 233 (1959).

[249] G. Turin, "An introduction to matched filters", IRE Transactions on Information Theory **6**, 311 (1960).

[250] C. Cook, *Radar Signals: An Introduction to Theory and Application*, 1st edition (Academic Press, 1967).

[251] B. P. Abbott et al., "Observation of Gravitational Waves from a Binary Black Hole Merger", Physical Review Letters **116**, 061102 (2016).

[252] F. Marsili, F. Najafi, E. Dauler, F. Bellei, X. Hu, M. Csete, R. J. Molnar, and K. K. Berggren, "Single-Photon Detectors Based on Ultranarrow Superconducting Nanowires", Nano Letters **11**, 2048 (2011).

[253] N. Fruitwala, P. Strader, G. Cancelo, T. Zmuda, K. Treptow, N. Wilcer, C. Stoughton, A. B. Walter, N. Zobrist, G. Collura, I. Lipartito, J. I. Bailey, and B. A. Mazin, "Second generation readout for large format photon counting microwave kinetic inductance detectors", Review of Scientific Instruments **91**, 124705 (2020).

[254] H. Hao, Q.-Y. Zhao, L.-D. Kong, S. Chen, H. Wang, Y.-H. Huang, J.-W. Guo, C. Wan, H. Liu, X.-C. Tu, L.-B. Zhang, X.-Q. Jia, J. Chen, L. Kang, C. Li, T. Chen, G.-X. Cao, and P.-H. Wu, "Improved pulse discrimination for a superconducting series nanowire detector by applying a digital matched filter", Applied Physics Letters **119**, 232601 (2021).

[255] P. Kiefer, V. Hahn, M. Nardi, L. Yang, E. Blasco, C. Barner-Kowollik, and M. Wegener, "Sensitive Photoresists for Rapid Multiphoton 3D Laser Micro- and Nanoprinting", Advanced Optical Materials **8**, 2000895 (2020).

[256] T. Baldacchini, editor, *Three-Dimensional Microfabrication Using Two-Photon Polymerization*, 2nd edition, Micro and Nano Technologies (William Andrew, 2019).

[257] M. Malinauskas, A. Žukauskas, S. Hasegawa, Y. Hayasaki, V. Mizeikis, R. Buividas, and S. Juodkazis, "Ultrafast laser processing of materials: from science to industry", Light: Science & Applications **5**, e16133 (2016).

[258] G. Cojoc, C. Liberale, P. Candeloro, F. Gentile, G. Das, F. De Angelis, and E. Di Fabrizio, "Optical micro-structures fabricated on top of optical fibers by means of two-photon photopolymerization", Microelectronic Engineering **87**, 876 (2010).

[259] Y. Xu, A. Kuzmin, E. Knehr, M. Blaicher, K. Ilin, P.-I. Dietrich, W. Freude, M. Siegel, and C. Koos, “Superconducting nanowire single-photon detector with 3D-printed free-form microlenses”, Optics Express **29**, 27708 (2021).

[260] N. Li, C. P. Ho, S. Zhu, Y. H. Fu, Y. Zhu, and L. Y. T. Lee, “Aluminium nitride integrated photonics: a review”, Nanophotonics **10**, 2347 (2021).

[261] C. E. Jimenez, “Fiber-coupled superconducting nanowire single-photon detector using photonic wirebonds”, master’s thesis (Karlsruhe Institute of Technology, 2020).

[262] S. Batsanov, E. Ruchkin, and I. Poroshina, *Refractive Indices of Solids*, 1st edition, SpringerBriefs in Applied Sciences and Technology (Springer, 2016).

[263] M. Blaicher, P.-I. Dietrich, and C. Koos, *Localization of optical coupling points*, patent WO2021175967A1 (2021).

[264] D. Meschede, *Optics, Light and Lasers: The Practical Approach to Modern Aspects of Photonics and Laser Physics*, 3rd edition (Wiley-VCH, 2017).

[265] M. Schmid, D. Ludescher, and H. Giessen, “Optical properties of photoresists for femtosecond 3D printing: refractive index, extinction, luminescence-dose dependence, aging, heat treatment and comparison between 1-photon and 2-photon exposure”, Optical Materials Express **9**, 4564 (2019).

[266] A. Kuzmin, S. Doerner, S. Singer, I. Charaev, K. Ilin, S. Wuensch, and M. Siegel, “Terahertz Transition-Edge Sensor With Kinetic-Inductance Amplifier at 4.2 K”, IEEE Transactions on Terahertz Science and Technology **8**, 622 (2018).

[267] J. R. Clem, B. Bumble, S. I. Raider, W. J. Gallagher, and Y. C. Shih, “Ambegaokar-Baratoff–Ginzburg-Landau crossover effects on the critical current density of granular superconductors”, Physical Review B **35**, 6637 (1987).

[268] Y. P. Gousev, G. N. Gol’tsman, A. D. Semenov, E. M. Gershenzon, R. S. Nebosis, M. A. Heusinger, and K. F. Renk, “Broadband ultrafast superconducting NbN detector for electromagnetic radiation”, Journal of Applied Physics **75**, 3695 (1994).

[269] F. C. Wellstood, C. Urbina, and J. Clarke, “Flicker (1/f) noise in the critical current of Josephson junctions at 0.09–4.2 K”, Applied Physics Letters **85**, 5296 (2004).

[270] J. Gao, J. Zmuidzinas, B. A. Mazin, H. G. LeDuc, and P. K. Day, “Noise properties of superconducting coplanar waveguide microwave resonators”, Applied Physics Letters **90**, 102507 (2007).

[271] G. G. Taylor, D. V. Morozov, C. T. Lennon, P. S. Barry, C. Sheagren, and R. H. Hadfield, “Infrared single-photon sensitivity in atomic layer deposited superconducting nanowires”, Applied Physics Letters **118**, 191106 (2021).

[272] T. Proslier, J. A. Klug, J. W. Elam, H. Claus, N. G. Becker, and M. J. Pellin, "Atomic Layer Deposition and Superconducting Properties of NbSi Films", The Journal of Physical Chemistry C **115**, 9477 (2011).

[273] J. A. Klug, T. Proslier, J. W. Elam, R. E. Cook, J. M. Hiller, H. Claus, N. G. Becker, and M. J. Pellin, "Atomic Layer Deposition of Amorphous Niobium Carbide-Based Thin Film Superconductors", The Journal of Physical Chemistry C **115**, 25063 (2011).

[274] T. Proslier, J. Klug, N. C. Becker, J. W. Elam, and M. Pellin, "Atomic Layer Deposition of Superconductors", ECS Transactions **41**, 237 (2011).

List of Own Publications

Peer-reviewed Journals

[209] E. Knehr, A. Kuzmin, D. Y. Vodolazov, M. Ziegler, S. Doerner, K. Ilin, M. Siegel, R. Stolz, and H. Schmidt, "Nanowire single-photon detectors made of atomic layer-deposited niobium nitride", Superconductor Science and Technology **32**, 125007 (2019).

[219] E. Knehr, M. Ziegler, S. Linzen, K. Ilin, P. Schanz, J. Plentz, M. Diegel, H. Schmidt, E. Il'ichev, and M. Siegel, "Wafer-level uniformity of atomic-layer-deposited niobium nitride thin films for quantum devices", Journal of Vacuum Science & Technology A **39**, 052401 (2021).

[241] E. Knehr, A. Kuzmin, S. Doerner, S. Wuensch, K. Ilin, H. Schmidt, and M. Siegel, "Synchronous single-photon detection with self-resetting, GHz-gated superconducting NbN nanowires", Applied Physics Letters **117**, 132602 (2020).

[259] Y. Xu, A. Kuzmin, E. Knehr, M. Blaicher, K. Ilin, P.-I. Dietrich, W. Freude, M. Siegel, and C. Koos, "Superconducting nanowire single-photon detector with 3D-printed free-form microlenses", Optics Express **29**, 27708 (2021).

Conferences

1. Emanuel Knehr, Yilin Xu, Yiyang Bao, Pascal Maier, Carlos Jimenez, Konstantin Ilin, Mario Ziegler, Heidemarie Schmidt, Sebastian Kempf, Artem Kuzmin, Christian Koos, Michael Siegel. Optical packaging of SNSPDs with photonic wire bonds. *15th European Conference on Applied Superconductivity, virtual, September 5 – September 9, 2021* (invited talk).

2. Emanuel Knehr, Artem Kuzmin, Steffen Doerner, Stefan Wuensch, Konstantin Ilin, Heidemarie Schmidt, Michael Siegel. GHz-gated operation and readout of microwave-biased superconducting nanowire detector arrays. *Applied Superconductivity Conference 2020, virtual, October 25 – November 6, 2020* (talk).

3. Artem Kuzmin, Emanuel Knehr, Yilin Xu, Steffen Doerner, Stefan Wuensch, Christian Koos, Michael Siegel. RF-SNSPD arrays: jitter and 3D laser-printed microlens coupling. *Applied Superconductivity Conference 2020, virtual, October 25 – November 6, 2020* (talk).

4. Artem Kuzmin, Steffen Doerner, Emanuel Knehr, Erik Heinz, Katja Peiselt, Stefan Wuensch, Michael Siegel. Membrane-isolated TES with microwave kinetic-inductance nanowire galvanometer. *Applied Superconductivity Conference 2020, virtual, October 25 – November 6, 2020* (poster).

5. Emanuel Knehr, Mario Ziegler, Sven Linzen, Artem Kuzmin, Konstantin Ilin, Ronny Stolz, Evgeni Ilichev, Heidemarie Schmidt, Michael Siegel. Atomic layer-deposited superconducting niobum nitride for quantum device applications. *AVS 20th International Conference on Atomic Layer Deposition, virtual, June 29 – July 1, 2020* (poster).

6. Emanuel Knehr, Julia Brandel, Oliver Brandel, Mario Ziegler, Sebastian Goerke, Sven Linzen, Uwe Hübner, Konstantin Ilin, Michael Siegel, Hans-Georg Meyer, Heidemarie Schmidt. Nanowire single-photon detectors made from atomic layer deposited NbN thin films. *683. WE-Heraeus-Seminar on "Physics and Applications of Superconducting Nanowire Single Photon Detectors", Bad Honnef, Germany, November 12 – November 16, 2018* (poster).

7. Emanuel Knehr, Julia Brandel, Oliver Brandel, Mario Ziegler, Sebastian Goerke, Sven Linzen, Uwe Hübner, Konstantin Ilin, Michael Siegel, Hans-Georg Meyer, Heidemarie Schmidt. Properties of NbN-based nanowire single-photon detectors fabricated by plasma-enhanced atomic layer deposition. *Applied Superconductivity Conference 2018, Seattle, USA, October 28 – November 2, 2018* (invited poster).

Danksagung

Die vorliegende Arbeit entstand während meiner Tätigkeit am Institut für Mikro- und Nanoelektronische Systeme (IMS) am Karlsruher Institut für Technologie (KIT), sowie am Leibniz-Institut für Photonische Technologien (Leibniz-IPHT) in Jena.

Zuallererst möchte ich meinem Doktorvater, Herrn Prof. Michael Siegel, für das Ermöglichen dieser Arbeit danken. Das große Vertrauen, das er mir schon seit der Bachelorarbeit entgegenbringt, und die Freiheit, die er mir hinsichtlich der Thematik gegeben hat, schätze ich sehr. Für die Förderung meiner Arbeit, auch während meiner Zeit zurück in Karlsruhe, danke ich Frau Prof. Heidemarie Schmidt. Des Weiteren danke ich Herrn Prof. Sebastian Kempf für den unkomplizierten Übergang nach seiner Übernahme als Institutsleitung am IMS und die weitere Unterstützung meiner Forschungsarbeit. Meinem Mentor und Freund Artem Kuzmin danke ich für die vielen gemeinsam entwickelten Ideen, die zahlreichen Diskussionen über Messergebnisse und Manuskripte und die große Geduld bei all meinen Fragen. Mit seinem Wissen, seiner Begeisterung und seiner unkonventionellen Herangehensweise an wissenschaftliche Problemstellungen hat er meine Arbeit sehr geprägt. Ein großen Anteil an der fachlichen Unterstützung hat auch Konstantin Ilin, der noch im Studium mit seinen Vorlesung mein Interesse an Supraleitung und Einzelphotonendetektoren geweckt hat und im Verlauf meiner Arbeit für jegliche Fragen theoretischer und technologischer Natur stets Zeit gefunden hat. Vielen Dank dafür. Ein besonderer Dank geht auch an Mario Ziegler. Er hat einen großen Teil dieser Arbeit möglich gemacht und ich danke ihm für seine offene Art, die Hilfe zu allen materialwissenschaftlichen Fragen und seinen großen Anstrengungen zur Herstellung der unzähligen Proben. Auch wenn meine Anwesenheit in Jena kürzer war als anfangs gedacht, war er immer erreichbar und die Zusammenarbeit reibungslos.

Durch die Tätigkeit an zwei Instituten hatte ich Kontakt zu vielen weiteren Kollegen, die mir an verschiedensten Stellen geholfen haben. Vielen Dank an Stefan Wünsch, Constantin Schuster, Michael Müller, Mathias Wegner und Patrick Schanz für die gute Zusammenarbeit im Institutsalltag. Alexander Stassen, Karlheinz Gutbrod, Frank Ruhnau, Alexander Hettinger und Stefan Ziegler danke ich für die technische Hilfe und Doris Duffner für die Unterstützung bei allem, was organisatorisch angefallen ist. Aus Jena danke ich vor allem Sven Linzen, Detlef Born und Jonathan Plentz für die Unterstützung bei den Messungen. Für die Zusammenarbeit in Bezug auf die optische Kopplung der Detektoren möchte ich mich bei Prof. Christian Koos, Yilin Xu, Yiyang Bao und Pascal Maier vom Institut für Photonik und Quantenelektronik (IPQ) am KIT, sowie dem Masterstudenten Carlos Jiménez bedanken.

Nicht zuletzt danke ich meiner Familie für die bedingungslose Unterstützung bei allem, was ich tue, und meiner Verlobten Jana für die vielen schönen Momente, die ich mit ihr im Leben abseits der Forschung erleben darf.

Karlsruher Schriftenreihe zur Supraleitung

Karlsruher Institut für Technologie (KIT) | ISSN 1869-1765

Band 001 **Christian Schacherer**
Theoretische und experimentelle Untersuchungen zur Entwicklung supraleitender resistiver Strombegrenzer.
ISBN 978-3-86644-412-6

Band 002 **Alexander Winkler**
Transient behaviour of ITER poloidal field coils.
ISBN 978-3-86644-595-6

Band 003 **André Berger**
Entwicklung supraleitender, strombegrenzender Transformatoren.
ISBN 978-3-86644-637-3

Band 004 **Christoph Kaiser**
High quality Nb/Al-AlOx/Nb Josephson junctions. Technological development and macroscopic quantum experiments.
ISBN 978-3-86644-651-9

Band 005 **Gerd Hammer**
Untersuchung der Eigenschaften von planaren Mikrowellen-resonatoren für Kinetic-Inductance Detektoren bei 4,2 K.
ISBN 978-3-86644-715-8

Band 006 **Olaf Mäder**
Simulationen und Experimente zum Stabilitätsverhalten von HTSL-Bandleitern.
ISBN 978-3-86644-868-1

Band 007 **Christian Barth**
High Temperature Superconductor Cable Concepts for Fusion Magnets.
ISBN 978-3-7315-0065-0

Band 008 **Axel Stockhausen**
Optimization of Hot-Electron Bolometers for THz Radiation.
ISBN 978-3-7315-0066-7

Band 009 **Petra Thoma**
Ultra-fast $YBa_2Cu_3O_{7-x}$ direct detectors for the THz frequency range.
ISBN 978-3-7315-0070-4

Band 010 **Dagmar Henrich**
Influence of Material and Geometry on the Performance of Superconducting Nanowire Single-Photon Detectors.
ISBN 978-3-7315-0092-6

Band 011 **Alexander Scheuring**
Ultrabreitbandige Strahlungseinkopplung in THz-Detektoren.
ISBN 978-3-7315-0102-2

Band 012 **Markus Rösch**
Development of lumped element kinetic inductance detectors for mm-wave astronomy at the IRAM 30 m telescope.
ISBN 978-3-7315-0110-7

Band 013 **Johannes Maximilian Meckbach**
Superconducting Multilayer Technology for Josephson Devices.
ISBN 978-3-7315-0122-0

Band 014 **Enrico Rizzo**
Simulations for the optimization of High Temperature Superconductor current leads for nuclear fusion applications.
ISBN 978-3-7315-0132-9

Band 015 **Philipp Krüger**
Optimisation of hysteretic losses in high-temperature superconducting wires.
ISBN 978-3-7315-0185-5

Band 016 **Matthias Hofherr**
Real-time imaging systems for superconducting nanowire single-photon detector arrays.
ISBN 978-3-7315-0229-6

Band 017 **Oliver Näckel**
Development of an Air Coil Superconducting Fault Current Limiter.
ISBN 978-3-7315-0526-6

Band 018 **Christoph M. Bayer**
Characterization of High Temperature Superconductor Cables for Magnet Toroidal Field Coils of the DEMO Fusion Power Plant.
ISBN 978-3-7315-0605-8

Band 019 **Shengnan Zou**
Magnetization of High Temperature Superconducting Trapped-Field Magnets.
ISBN 978-3-7315-0715-4

Band 020 **Ilya Charaev**
Improving the Spectral Bandwidth of Superconducting Nanowire Single-Photon Detectors (SNSPDs).
ISBN 978-3-7315-0745-1

Band 021 **Juliane Raasch**
Electrical-field sensitive $YBa_2Cu_3O_{7-x}$ detectors for real-time monitoring of picosecond THz pulses.
ISBN 978-3-7315-0786-4

Band 022 **Yingzhen Liu**
Design of a superconducting DC wind generator.
ISBN 978-3-7315-0796-3

Band 023 **Sebastian Hellmann**
Research and Technology Development on Superconducting Current Limiting Transformers.
ISBN 978-3-7315-0804-5

Band 024 **Simon J. Otten**
Characterisation of REBCO Roebel cables.
ISBN 978-3-7315-0904-2

Band 025 **Julia Brandel**
Supraleitende Einzelphotonenzähler: Optimierung der Zeitauflösung und Anwendungsbeispiele aus der Spektroskopie.
ISBN 978-3-7315-0917-2

Band 026 **Dustin Kottonau, Eugen Shabagin, Wescley T. B. de Sousa, Jörn Geisbüsch, Mathias Noe, Hanno Stagge, Simon Fechner, Hannes Woiton, Thomas Küsters**
Bewertung des Einsatzes supraleitender 380-kV-Kabel.
ISBN 978-3-7315-0927-1

Band 027 **Steffen Dörner**
Multifrequenzausleseverfahren von supraleitenden Einzelphotonen-Detektoren.
ISBN 978-3-7315-0961-5

Band 028 **Michael Merker**
Superconducting integrated THz receiver.
ISBN 978-3-7315-0970-7

Band 029 **Wolfgang-Gustav Ekkehart Schmidt**
Superconducting Nanowire Single-Photon Detectors for Quantum Photonic Integrated Circuits on GaAs.
ISBN 978-3-7315-0980-6

Band 030 **Dustin Kottonau, Eugen Shabagin, Wescley de Sousa, Jörn Geisbüsch, Mathias Noe, Hanno Stagge, Simon Fechner, Hannes Woiton, Thomas Küsters**
Evaluation of the Use of Superconducting 380 kV Cable.
ISBN 978-3-7315-1026-0

Band 031 **Alan Preuß**
Development of high-temperature superconductor cables for high direct current applications.
ISBN 978-3-7315-1041-3

Band 032 **Roland Gyuráki**
Fluorescent thermal imaging method for investigating transient effects in high-temperature superconductor tapes and coils.
ISBN 978-3-7315-1064-2

Band 033 **Aurélien Godfrin**
AC Losses in High-Temperature Superconductor Tapes and Cables for Power Applications.
ISBN 978-3-7315-1096-3

Band 034 **Alexander Schmid**
Integrierte planare Antennen für supraleitende Detektorsysteme zur THz-Strahldiagnostik.
ISBN 978-3-7315-1145-8

Band 035 **Alexander Buchholz**
Prospective Life Cycle Assessment of High-Temperature Superconductors for Future Grid Applications.
ISBN 978-3-7315-1194-6

Band 036 **Emanuel Marius Knehr**
Technology and readout for scaling up superconducting nanowire single-photon detectors.
ISBN 978-3-7315-1256-1